Applications of Pattern Recognition

Edited by Carlos M. Travieso-Gonzalez

Published in London, United Kingdom

IntechOpen

Supporting open minds since 2005

Applications of Pattern Recognition
http://dx.doi.org/10.5772/intechopen.80151
Edited by Carlos M. Travieso-Gonzalez

Contributors
Nwagwu Honour Chika Chika, Emmanuel Ukekwe, Celestine Ugwoke, Dora Ndjinkou Ndoumbe, George Okereke, Roi Santos, Xose M. Pardo, Xose Ramón Fdez-Vidal, Bo-Wei Chen, Jia-Ching Wang, Ismail Jamail, Ahmed Moussa, Lukáš Semerád, Martin Drahansky, Anwar Ebrahim, Hoshang Kolivand, Norio Tagawa

First published in London, United Kingdom, 2021 by IntechOpen
IntechOpen is the global imprint of INTECHOPEN LIMITED, registered in England and Wales, registration number: 11086078, 5 Princes Gate Court, London, SW7 2QJ, United Kingdom
Printed in Croatia

British Library Cataloguing-in-Publication Data
A catalogue record for this book is available from the British Library

Additional hard and PDF copies can be obtained from orders@intechopen.com

Applications of Pattern Recognition
Edited by Carlos M. Travieso-Gonzalez
p. cm.
Print ISBN 978-1-78985-332-2
Online ISBN 978-1-78985-561-6
eBook (PDF) ISBN 978-1-83880-913-3

Meet the editor

Carlos M. Travieso-González received his MSc degree in Telecommunication Engineering at Polytechnic University of Catalonia (UPC), Spain in 1997, and his Ph.D. degree in 2002 at the University of Las Palmas de Gran Canaria (ULPGC-Spain). He is a full professor of signal processing and pattern recognition and is head of the Signals and Communications Department at ULPGC, teaching from 2001 on subjects on signal processing and learning theory. His research lines are biometrics, biomedical signals and images, data mining, classification system, signal and image processing, machine learning, and environmental intelligence. He has researched in 52 international and Spanish research projects, some of them as head researcher. He is co-author of 4 books, co-editor of 27 proceedings books, guest editor for 8 JCR-ISI international journals, and up to 24 book chapters. He has over 450 papers published in international journals and conferences (81 of them indexed on JCR – ISI - Web of Science). He has published seven patents in the Spanish Patent and Trademark Office. He has been a supervisor on 8 Ph.D. theses (11 more are under supervision), and 130 master theses. He is the founder of The IEEE IWOBI conference series and the president of its Steering Committee, as well as the founder of both the InnoEducaTIC and APPIS conference series. He is an evaluator of project proposals for the European Union (H2020), Medical Research Council (MRC, UK), Spanish Government (ANECA, Spain), Research National Agency (ANR, France), DAAD (Germany), Argentinian Government, and the Colombian Institutions. He has been a reviewer in different indexed international journals (<70) and conferences (<250) since 2001. He has been a member of the IASTED Technical Committee on Image Processing from 2007 and a member of the IASTED Technical Committee on Artificial Intelligence and Expert Systems from 2011.

He has held the general chair position for the following: ACM-APPIS (2020, 2021), IEEE-IWOBI (2019, 2020 and 2020), A PPIS (2018, 2019), IEEE-IWOBI (2014, 2015, 2017, 2018), InnoEducaTIC (2014, 2017), IEEE-INES (2013), NoLISP (2011), JRBP (2012), and IEEE-ICCST (2005)

He is an associate editor of the Computational Intelligence and Neuroscience Journal (Hindawi – Q2 JCR-ISI). He was vice dean from 2004 to 2010 in the Higher Technical School of Telecommunication Engineers at ULPGC and the vice dean of Graduate and Postgraduate Studies from March 2013 to November 2017. He won the "Catedra Telefonica" Awards in Modality of Knowledge Transfer, 2017, 2018, and 2019 editions, and awards in Modality of COVID Research in 2020.

Public References:

Researcher ID	http://www.researcherid.com/rid/N-5967-2014
ORCID	https://orcid.org/0000-0002-4621-2768
Scopus Author ID	https://www.scopus.com/authid/detail.uri?authorId=6602376272
Scholar Google	https://scholar.google.es/citations?user=G1ks9nIAAAAJ&hl=en
ResearchGate	https://www.researchgate.net/profile/Carlos_Travieso

Contents

Preface

Pattern recognition is an important field during this time, due to its use in companies, business, and real life. Each time more and more data are generated, there are more identifications of patterns. The automatic interpretation of big data is by the extraction of patterns, and the field of pattern recognition has a great role in extracting information and decision-making. It is key to the development of a society and our quality of life, based on the knowledge gained from this data.

This book will show various applications for the improvement and use of data analysis and the automatic system available as tools for security, biology, and biometrics.

In fields with the use of security, molecular biology, modeling, improvement of data, and biometrics, new and advanced techniques can be applied to facilitate or provide tools for the detection of patterns. A great number of tools are being developed in this sense. This book presents works of high quality, developed on a scientific methodology, giving validation to the present proposals. The content of the book focuses on automatic systems and data analysis; therefore, it will be a very attractive read for the reader.

"Applications of Pattern Recognition" comprises seven chapters, which have been divided into two sections: "Data Analysis" and "Automatic Systems". The section "Data Analysis" has three chapters. The analysis is applied on an approach that is composed of two types of depth restoration methods based on fixation tremor: differential type method and integral type method. The first method is based on the change in image brightness between frames, and the other type is based on image blurring due to movement. In this section, an approach is focused on the discussion of identifying inconsistencies associated with patterns. And finally, a 3D abstraction method receives input from camera intrinsic parameters and several pictures of the scene. This approach introduces the geometrical relations, which have to be exploited for structure from motion sketch or abstraction based on line segments, the optimization methods for its optimization, and how to compare the experimental results with ground truth measurements. The section "Automatic Systems" contains four chapters. The automatic systems are shown by an approach, which introduces recent methods for processing missing values. For this approach, four types of commonly used algorithms were applied, namely, k-nearest neighbors, regression, tree-based algorithms, and latent component-based approaches. This book also presents a system based on different approaches using the retina of the human eye to evaluate individual parameters for human recognition. Furthermore, a method is developed for the feature selection using discrete cosine transform to extract the feature, and then, the sparse principal component analysis is used for the selection of significant attributes, as the feature technique in offline Arabic signature verification. Finally, clustering algorithms and statistical approaches are shown for grouping similar gene expression profiles that can be applied to RNA-seq data analysis.

As editor of this book, I would like to thank the authors for their effort and dedication that they have made to achieve some works of great quality. The sum of this effort has produced this book, which has become an inescapable read for all those who want to know the latest advances in pattern recognition.

Carlos M. Travieso-Gonzalez
Signals and Communications Department (DSC),
Institute for Technological Development and Innovation in Communications (IDeTIC),
University of Las Palmas de Gran Canaria (ULPGC),
Las Palmas de Gran Canaria,
Spain

Section 1

Data Analysis

Chapter 1

Stereoscopic Calculation Model Based on Fixational Eye Movements

Norio Tagawa

Abstract

Fixational eye movement is an essential function for watching things using the retina, which has the property of responding only to changes in incident light. However, since the rotation of the eyeball causes the translational movement of the crystalline lens, it is possible in principle to recover the depth of the object from the moving image obtained in this way. We have proposed two types of depth restoration methods based on fixation tremor; differential-type method and integral-type method. The first is based on the change in image brightness between frames, and the latter is based on image blurring due to movement. In this chapter, we introduce them and explain the simulations and experiments performed to verify their operation.

Keywords: motion stereoscopic, fixational eye movements, differential-type method, integral-type method, optical flow, gradient equation, image blur

1. Introduction

When humans stare at a target, an irregular involuntary movement called fixational eye movements occur [1]. The human retina can maintain reception sensitivity by finely vibrating the image of the target on the retina, so in order to see something, first fixation motion is required. It has been reported that the vibrations may work not only as such the intrinsic function to preserve photosensitivity but also as an assistance in image analysis, the mechanism of which can be interpreted as an instance of stochastic resonance (SR) [1]. SR is inspired by biology, more specifically by neuron dynamics [2], and based on it, the Dynamic Retina (DR) [3] and the Resonant Retina (RR) [4], which are new vision devices taking advantage of random camera vibrations, were proposed for contrast enhancement and edge detection respectively. It has been reported that the movement of the retinal image due to fixation eye movements can be an unconscious clue to depth perception, and an actual vision system based on fixational eye movements has been proposed [5].

On the other hand, binocular stereopsis is vigorous and plays an essential role in depth perception of a human vision system [6]. In general, binocular stereopsis detects relatively large disparities, hence it can recognize high accurate depth. However this causes an occlusion problem, and a lot of solutions of it have been proposed. Wang et al. have proposed a local detector for occlusion based on deep learning [7]. In [8], a robust depth restoration method has been proposed that integrates line-field imaging technology that simultaneously observes multiple angle views with stereo vision. Therefore, we expect that primitive depth information detected by fixational eye movements can be used to solve occlusion for binocular stereopsis. There is a concern that the accuracy of depth restoration by a small camera motion is lower than that of stereo vision. Even so, it is expected that erroneous correspondence due to the existence of occlusion can be reduced by using the depth information from fixational eye movements for the correspondence problem in stereo vision.

In monocular stereoscopic vision, "structure from motion (SFM)" has been the most widely studied, and many remarkable results have been reported. SFM has various calculation principles. To achieve spatially dense depth recovery with high computational efficiency, a method based on the gradient equation that expresses the constraint between the spatiotemporal derivative values of image intensity and the movement on the image is effective [9–11]. It should be noted that for such a gradient method, there is an appropriate size of movement to recover the correct depth. Since the gradient equation holds perfectly for small motions, the error in the equation cannot be ignored for very large motions. On the contrary, in the case of small movement, the motion information is buried in the observation error of the spatiotemporal derivative in intensity.

Adaptation of the frame rate is required to make the motion size suitable for the gradient method. We have proposed a method that does not require a variable frame rate based on multi-resolution decomposition of images, but it requires high computational cost [12]. Therefore, we focus on small movements with an emphasis on avoiding equation errors in the gradient method. Then, in order to solve the above signal-to-noise ratio (S/N) problem that occurs with small movements, many observations are collected and used all at once [13, 14]. In such a strategy, it is desirable that the direction and size of the motion take different values. From the above discussion, we examined a depth perception model based on fixational eye movement and gradient method. Fixational eye movements are divided into three types: microsaccades, drifts, and tremors. As the first report of our attempt, we focused on tremor, the smallest of the three types. In the next step, we plan to use drift and microsaccade analogies for further progress. Using a lot of images captured with random small motions of camera, which consists of three-dimensional (3-D) rotations imitating fixational eyeball motions [1], many observations can be used at each pixel, i.e. many gradient equations can be used to recover the each depth value corresponding to the each pixel. Since the difference between the center of the three-dimensional rotation and the lens center generates a translational motion of the lens center, depth information can be obtained from these images. Simulations with artificial images confirm that the proposed method works effectively when the observed noise is an actual sample of a theoretically defined noise model.

However, if the wavelength of the main luminance pattern is small compared to the size of the motion in the image, aliasing will occur and the gradient equation will be useless. In other words, the methods of [13, 14] cannot be applied. To avoid the above problem, we proposed a new scheme based on the integral form that also used the analogy of fixational eye movement [15, 16]. Add up the many images generated by the above method to get one blurry image. The degree of blur is a function of pixel position and also depends on the depth value of each pixel. That is, the difference in the degree of image blur indicates the depth information. Based on the proposed scheme, the spatial distribution of the image blur is effectively estimated using the blurred image and the original image without blur. By modeling the small 3-D rotation of the camera as a Gaussian random variable, the depth map can be calculated analytically from this blur distribution.

Several depth recovery methods using motion-blur have been already proposed, but those use the blur caused by definite and simple camera motions. For example, blur by a translational camera motion is used in [17], and blur by an unconstrained camera motion composed of translation and rotation is assumed in [18]. The depth recovery performance of these methods depends on the orientation of the texture in the image. That is, if the texture has a strip pattern whose direction is parallel to the direction of motion in the image, there is less blurring and accurate depth recovery is difficult. Unlike such a specific camera motion, the random camera rotation used in this study works well for any texture. Deterministic camera motion can be used just to solve this problem, but since it does not require precise control of the camera, it is easy to implement random camera rotation in a real system.

We proposed two algorithms based on the integral scheme. The first algorithm detects a point spread function (PSF) that represents image blur and then analyzes it to restore depth [15]. The second algorithm directly calculates the depth without detecting the PSF [16]. These algorithms use a motion-blurred image and a reference un-blurred image. It is expected that the performance of the proposed scheme depends on the degree of motion blur. For the same PSF, i.e. the fixed deviation of the random camera rotations, fine texture is advantageous for observing the accurate blur. This characteristic is the opposite of the method based on the differential scheme based on the gradient equation.

The features of our methods described above can be summarized as follows.

1. In a camera motion model that simulates eyeball rotation, the translation of the lens center is secondarily generated by eye rotation, reducing the number of unknown parameters. This camera motion model also allows you to recover depth as an absolute quantity instead of a relative quantity.

2. The movement of the camera is subtle because it simulates the tremor component of the eyeball. Therefore, by using a large number of image pairs, it is possible to improve the accuracy of depth recovery while avoiding occlusion.

3. In order to use multiple image pairs at the same time, we have adopted a direct framework that explicitly uses the inverse depth, which is a common parameter for them.

In the following, we will first explain the camera model and the imaging system, and then explain the differential scheme and the integral scheme, and the algorithms based on each. Due to the limitation of the number of pages, the integer type explains only the direct method. The function and characteristics of each algorithm are shown as the result of computer simulation with an emphasis on quantitative comparison with the true value. Finally, one of the algorithms in the differential scheme is applied to the real image, and the result is explained.

2. Camera motions imitating tremor and projection model

In this research, we use a perspective projection system as a camera imaging model. The camera is fixed in the (X, Y, Z) coordinate system. The lens center corresponding to the viewpoint is at the origin O, and the optical axis is along the Z axis. By taking the focal length as a unit of geometric representation, we can assume image plane $Z = 1$ without loss of generality. The point $(X, Y, Z)^{\mathrm{T}}$ on the object in 3-D space is projected on the image point $\boldsymbol{x} \equiv (x, y, 1)^{\mathrm{T}} = (X/Z, Y/Z, 1)^{\mathrm{T}}$.

A brief description of the camera motion model that mimics the tremor component of fixational eye movements proposed in the previous study [12]. According to the analogy of the human eyeball, the center of camera rotation is set behind the lens center by Z_0 along the optical axis, and there is no explicit translational motion of the camera. This rotation vector $\boldsymbol{r} = (r_x, r_y, r_z)^{\mathrm{T}}$ also expresses a rotation vector centered on the coordinate origin, which is the lens center, using the same component. On the other hand, this difference between the coordinate origin and the center of rotation results in the translation vector $\boldsymbol{u} = (u_x, u_y, u_z)^{\mathrm{T}}$, which is formulated as follows:

$$\begin{bmatrix} u_x \\ u_y \\ u_z \end{bmatrix} = \begin{bmatrix} r_x \\ r_y \\ r_z \end{bmatrix} \times \begin{bmatrix} 0 \\ 0 \\ Z_0 \end{bmatrix} = Z_0 \begin{bmatrix} r_y \\ -r_x \\ 0 \end{bmatrix}. \tag{1}$$

In general, the translational motion of a camera lens is essential to recover depth, and our camera motion model can implicitly achieve that translation simply by rotating the camera. This facilitates camera control. In addition, the system can recover absolute depth by pre-calibrating Z_0. We show the coordinate system and camera motion model used in this study in **Figure 1**.

From Eq. (1), it can be known that r_z causes no translations. Therefore, we can set $r_z = 0$ and redefine $\boldsymbol{r} = (r_x, r_y, 0)^{\mathrm{T}}$ as a rotational vector like an eyeball. Using Eq. (1) and the inverse depth $d(x, y) = 1/Z(x, y)$, image motion called "optical flow" $\mathbf{v} = [v_x, v_y]^{\mathrm{T}}$ is given as follows:

$$v_x = xyr_x - (1 + x^2)r_y - Z_0 r_y d, \tag{2}$$

$$v_y = (1 + y^2)r_x - xyr_y + Z_0 r_x d. \tag{3}$$

In the above equations, d is an unknown variable at each pixel, and $\boldsymbol{u}$ and $\boldsymbol{r}$ are unknown common parameters for the whole image.

In this study, we treat $\boldsymbol{r}(t)$ as a white stochastic process to simplify the motion model, and t indicates time. $\boldsymbol{r}(t)$ is defined as the rotation speed with respect to the camera orientation of $t = 0$, not the derivative between consecutive frames. In the actual fixational eye movement, the temporal correlation of tremor that forms the drift component is ignored. We assume that each fluctuation of $\boldsymbol{r}(t)$

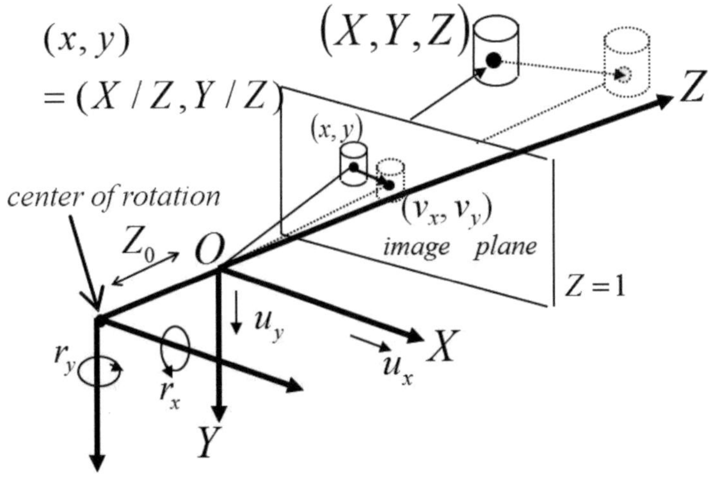

Figure 1.
Coordinate system and camera motion model used in this study.

follows a two-dimensional (2-D) Gaussian distribution with a mean of $\mathbf{0}$ and a variance–covariance matrix of $\sigma_r^2 \boldsymbol{I}$ with an identity matrix $\boldsymbol{I}$.

$$p\left(\boldsymbol{r}(t)|\sigma_r^2\right) = \frac{1}{\left(\sqrt{2\pi}\sigma_r\right)^2} \exp\left\{-\frac{\boldsymbol{r}(t)^{\mathrm{T}}\boldsymbol{r}(t)}{2\sigma_r^2}\right\}. \tag{4}$$

In this study, $\boldsymbol{r}$ is defined as the rotational speed for ease of theoretical analysis. In a real system, we have no choice but to use finite rotation, but if the angle of rotation value is small, then the formulations are almost valid.

3. Differential-type method

3.1 Gradient equation for rigid motion

The general gradient equation is the first approximation of the assumption that image brightness is invariable before and after the relative 3-D motion between a camera and an object. Assuming that the brightness values before and after 3-D motion are equal, the image brightness after 3-D motion are expressed by Taylor expansion, and terms of degree 2 and above are ignored. As a result, at each pixel (x, y), the gradient equation is formulated with the partial differentials f_x, f_y and f_t, where t denotes time, of the image brightness $f(x, y, t)$ and the optical flow, as follows:

$$f_t = -f_x v_x - f_y v_y. \tag{5}$$

By substituting Eqs. (2) and (3) into Eq. (5), the gradient equation representing a rigid motion constraint can be derived explicitly.

$$f_t = -\left(f_x v_x^r + f_y v_y^r\right) - \left(-f_x r_y + f_y r_x\right) Z_0 d \equiv -f^r - f^u d. \tag{6}$$

3.2 Probabilistic model for differential-type method

Let M be the number of pairs of two frames and N be the number of pixels. In our study, $\left\{f_t^{(i,j)}\right\}_{i=1,\cdots,N; j=1,\cdots,M}$ and $\left\{\boldsymbol{r}^{(j)}\right\}_{j=1,\cdots,M}$ are treated as random variables, and $\left\{d^{(i)}\right\}_{i=1,\cdots,N}$ corresponding to the inverse depth of each pixel is treated as stochastic variable and recovered individually for each pixel. However, multiple frames $\left\{\boldsymbol{r}^{(j)}\right\}$ that vibrate due to irregular rotation are used for processing, but no pixel tracking is done. Therefore, the recovered $d^{(i)}$ at each pixel does not correspond exactly to the value of this pixel, but takes the mean of the adjacent regions defined by the vibration width of the image. As a result, the recovered $d^{(i)}$ correlates with the values in the adjacent regions. Therefore, from the beginning, $d^{(i)}$ should be treated as a variable with such a correlation. Based on tremor, $d^{(i)}$, which is estimated to correlate with the neighborhood, is planned to be improved to d for each pixel when dealing with drift and microsaccade in future research.

In this study, we assume that optical flow is very small, and hence, observation errors of f_t, f_x and f_y, which are calculated by finite difference, are small. Additionally, equation error is also small, and therefore we can assume that error having no relation with f_t, f_x and f_y is added to the whole gradient equation. From this consideration, we assume that $f_t^{(i,j)}$ is a Gaussian random variable with mean 0 and variance σ_o^2, and $f_x^{(i,j)}$ and $f_y^{(i,j)}$ have no error.

$$p\left(f_t^{(i,j)}|d^{(i)}, \boldsymbol{r}^{(j)}, \sigma_o^2\right) = \frac{1}{\sqrt{2\pi}\sigma_o} \times \exp\left\{-\frac{\left(f_t^{(i,j)} + f^{r(i,j)} + f^{u(i,j)}d^{(i)}\right)^2}{2\sigma_o^2}\right\}, \tag{7}$$

where $i = 1, \cdots, N$ and $j = 1, \cdots, M$, and σ_o^2 is an unknown variance.

As mentioned above, considering the neighborhood correlation of d to be recovered, in this study, to simplify modeling, we use the following equation as the depth model.

$$p\left(\boldsymbol{d}|\sigma_d^2\right) = \frac{1}{\left(\sqrt{2\pi}\sigma_d\right)^N} \exp\left\{-\frac{\boldsymbol{d}^{\mathrm{T}}\boldsymbol{L}\boldsymbol{d}}{2\sigma_d^2}\right\}, \tag{8}$$

where $\boldsymbol{d}$ is a N-dimensional vector composed of $\left\{d^{(i)}\right\}$ and $\boldsymbol{L}$ indicates a matrix corresponding to the 2-D Laplacian operator with a free end condition. By assuming this probabilistic density, we make a recovered depth map smooth. In this study, the variance σ_d^2 is controled heuristically in consideration of smoothness of a recovered depth map. In future, we are going to examine a strategy for determination of σ_d^2 in the whole system which models all fixational movements including microsaccade and drift also. Hereafter, we use the definition $\Theta \equiv \{\sigma_o^2, \sigma_r^2\}$. In the simulation described later, random rotation is sampled according to Eq. (4) as the rotation for the initial image, but since the rotation between successive frames is estimated during depth restoration, the determined σ_r^2 and the set value are different.

3.3 Algorithm for differential-type method

By applying the MAP-EM algorithm [19], parameter $\{\boldsymbol{d}, \Theta\}$ can be estimated as a MAP estimator based on $p\left(\boldsymbol{d}, \Theta|\left\{f_t^{(i,j)}\right\}\right)$, which is formulated by marginalizing the joint probability $p\left(\left\{\boldsymbol{r}^{(j)}\right\}, \boldsymbol{d}, \Theta|\left\{f_t^{(i,j)}\right\}\right)$ with respect to $\left\{\boldsymbol{r}^{(j)}\right\}$, but the prior of Θ is formally regarded as an uniform distribution. Additionally, $\left\{\boldsymbol{r}^{(j)}\right\}$ can be estimated as a MAP estimator based on $p\left(\left\{\boldsymbol{r}^{(j)}\right\}|\left\{f_t^{(i,j)}\right\}, \hat{\Theta}, \hat{\boldsymbol{d}}\right)$, in which $\hat{}$ means a MAP estimator described above.

In the EM scheme, $\left\{\left\{f_t^{(i,j)}\right\}, \left\{\boldsymbol{r}^{(j)}\right\}\right\}$ is considered as a complete data, $\left\{\boldsymbol{r}^{(j)}\right\}$ is treated as a missing data, and $\{\boldsymbol{d}, \Theta\}$ is treated as an unknown parameter. E step and M step are mutually repeated until they converge. At first, at the E step, calculate the conditional expectation of the log likelihood of the complete data with observing $\left\{f_t^{(i,j)}\right\}$, using the current MAP estimates $\left\{\hat{\boldsymbol{d}}, \hat{\Theta}\right\}$. It is generally called Q function. Especially for the MAP-EM algorithm, the objective function $J(\boldsymbol{d}, \Theta)$ maximized at the M step is equal to the Q function augmented by the log prior densities of parameters. In the following, the values computed using $\hat{\Theta}$ are indicated as $\hat{}$ also.

Based on the densities defined at 3.2, the objective function is derived as

$$
\begin{aligned}
J(\boldsymbol{d}, \Theta) = &\text{Const.} - \frac{MN}{2}\ln\sigma_o^2 - \frac{3M}{2}\ln\sigma_r^2 \\
&-\frac{1}{2\sigma_o^2}\sum_{j=1}^{M}\left\{\sum_{i=1}^{N} f_t^{(i,j)2} + 2\left(\sum_{i=1}^{N} f_t^{(i,j)}\boldsymbol{w}^{(i,j)\mathrm{T}}\right)\hat{\boldsymbol{r}}_m^{(j)} + \mathrm{tr}\left[\left(\sum_{i=1}^{N}\boldsymbol{w}^{(i,j)}\boldsymbol{w}^{(i,j)\mathrm{T}}\right)\hat{\boldsymbol{R}}^{(j)}\right]\right\} \\
&-\frac{1}{2\sigma_r^2}\sum_{j=1}^{M}\mathrm{tr}\hat{\boldsymbol{R}}^{(j)} - \frac{\boldsymbol{d}^{\mathrm{T}}\boldsymbol{L}\boldsymbol{d}}{2\sigma_d^2},
\end{aligned}
\tag{9}
$$

using the following definitions derived by formulating the posterior density $p\left(\{\boldsymbol{r}^{(j)}\}|\{f_t^{(i,j)}\}, \hat{\boldsymbol{d}}, \hat{\Theta}\right)$.

$$\hat{\boldsymbol{r}}_m^{(j)} \equiv \mathrm{E}\left[\boldsymbol{r}^{(j)}|\{f_t^{(i,j)}\}, \hat{\boldsymbol{d}}, \hat{\Theta}\right] = -\frac{1}{\hat{\sigma}_o^2}\hat{\boldsymbol{V}}_r^{(j)}\sum_{i=1}^{N} f_t^{(i,j)}\hat{\boldsymbol{w}}^{(i,j)}, \tag{10}$$

$$\hat{\boldsymbol{V}}_r^{(j)} = \left(\frac{1}{\hat{\sigma}_o^2}\sum_{i=1}^{N}\hat{\boldsymbol{w}}^{(i,j)}\hat{\boldsymbol{w}}^{(i,j)\mathrm{T}} + \frac{1}{\hat{\sigma}_r^2}\boldsymbol{I}\right)^{-1}, \tag{11}$$

$$\hat{\boldsymbol{R}}^{(j)} \equiv \mathrm{E}\left[\boldsymbol{r}^{(j)}\boldsymbol{r}^{(j)\mathrm{T}}|\{f_t^{(i,j)}\}, \hat{\boldsymbol{d}}, \hat{\Theta}\right] = \hat{\boldsymbol{V}}_r^{(j)} + \hat{\boldsymbol{r}}_m^{(j)}\hat{\boldsymbol{r}}_m^{(j)\mathrm{T}}, \tag{12}$$

$$\boldsymbol{w}^{(i,j)} = \begin{pmatrix} f_x^{(i,j)}x^{(i)}y^{(i)} + f_y^{(i,j)}\left(1+y^{(i)2}\right) \\ -f_x^{(i,j)}\left(1+x^{(i)2}\right) - f_y^{(i,j)}x^{(i)}y^{(i)} \end{pmatrix} + Z_0 d^{(i)}\begin{pmatrix} f_y^{(i,j)} \\ -f_x^{(i,j)} \end{pmatrix} \equiv \boldsymbol{w}_0^{(i,j)} + Z_0 d^{(i)}\boldsymbol{w}_d^{(i,j)}. \tag{13}$$

In the M step, $\{\boldsymbol{d}, \Theta\}$ are updated so that Eq. (9) is maximized. Eq. (9) can be rewritten as follows, ignoring constant values.

$$J(\boldsymbol{d}, \Theta) = -\frac{MN}{2}\ln\sigma_o^2 - M\ln\sigma_r^2 - \frac{1}{2\sigma_o^2}\hat{\mathrm{F}}\left(\{d^{(i)}\}\right) - \frac{1}{2\sigma_r^2}\hat{\mathrm{G}} - \frac{\boldsymbol{d}^{\mathrm{T}}\boldsymbol{L}\boldsymbol{d}}{2\sigma_d^2}. \tag{14}$$

From this representation, σ_o^2 and σ_r^2 can be updated as

$$\sigma_o^2 = \frac{\hat{\mathrm{F}}\left(\{d^{(i)}\}\right)}{MN}, \quad \sigma_r^2 = \frac{\hat{\mathrm{G}}}{2M}. \tag{15}$$

For $\boldsymbol{d}$, the partial derivative for each $d^{(i)}$ in the last term of Eq. (9) contains the adjacent surrounding ds. Therefore, to update $\boldsymbol{d}$, we need to solve the simultaneous equations. To avoid this, use the One-Step-Late (OSL) method [20]. That is, consider the surrounding d s as a constant and evaluate it with the current estimate $\hat{\boldsymbol{d}}$s. This allows each $d^{(i)}$ to be updated individually as follows:

$$
\begin{aligned}
d^{(i)} = &\frac{\hat{\sigma}_o^2\hat{\bar{d}}^{(i)}}{\sigma_d^2 Z_0^2\sum_{j=1}^{M}\mathrm{tr}\left(\boldsymbol{A}^{(i,j)}\hat{\boldsymbol{R}}^{(j)}\right) + \hat{\sigma}_o^2} \\
&-\frac{\sigma_d^2 Z_0\sum_{j=1}^{M}\left\{f_t^{(i,j)}\boldsymbol{w}_d^{(i,j)\mathrm{T}}\hat{\boldsymbol{r}}_m^{(j)} + \mathrm{tr}\left(\boldsymbol{B}^{(i,j)}\hat{\boldsymbol{R}}^{(j)}\right)\right\}}{\sigma_d^2 Z_0^2\sum_{j=1}^{M}\mathrm{tr}\left(\boldsymbol{A}^{(i,j)}\hat{\boldsymbol{R}}^{(j)}\right) + \hat{\sigma}_o^2},
\end{aligned}
\tag{16}
$$

where σ_o^2 is also evaluated by the current estimate, and the matrices $\boldsymbol{A}^{(i,j)}$ and $\boldsymbol{B}^{(i,j)}$ are defined as

$$\boldsymbol{A}^{(i,j)} \equiv \boldsymbol{w}_d^{(i,j)} \boldsymbol{w}_d^{(i,j)\mathrm{T}}, \tag{17}$$

$$\boldsymbol{B}^{(i,j)} \equiv \left(\boldsymbol{w}_d^{(i,j)} \boldsymbol{w}_0^{(i,j)\mathrm{T}} + \boldsymbol{w}_0^{(i,j)} \boldsymbol{w}_d^{(i,j)\mathrm{T}} \right)/2, \tag{18}$$

and $\overline{d}^{(i)}$ indicates the local mean using a four-neighboring system that does not include $d^{(i)}$.

3.4 Numerical experiments of differential-type method

In order to confirm the function of the proposed method, we conducted numerical experiments using artificial images. **Figure 2(a)** shows the original image generated using the depth map shown in **Figure 2(b)**. The image size is 128 × 128 pixels, which is equivalent to $-0.5 \leq x, y \leq 0.5$ measured in focal length units. In **Figure 2(b)**, the vertical axis shows the depth Z in units of focal length, and the horizontal axis shows the pixel position on the image plane.

In our model, the successive image pairs are used in turn to calculate f_t. This study ignores the drift component of fixation eye movements and assumes that there is no temporal divergence of the range of motion at each image position. Therefore, each rotation sampled as a Gaussian independent random variable is considered to be relative to the initial image shown in **Figure 2(a)**. We can think of a gradient equation that holds between the resulting image and the initial image. Additionally, in order to firstly justify our algorithm for the assumed statistical models, we computed $\{f_t\}$ using Eq. (6) with true values of $\boldsymbol{r}$ and $\{d\}$ and use

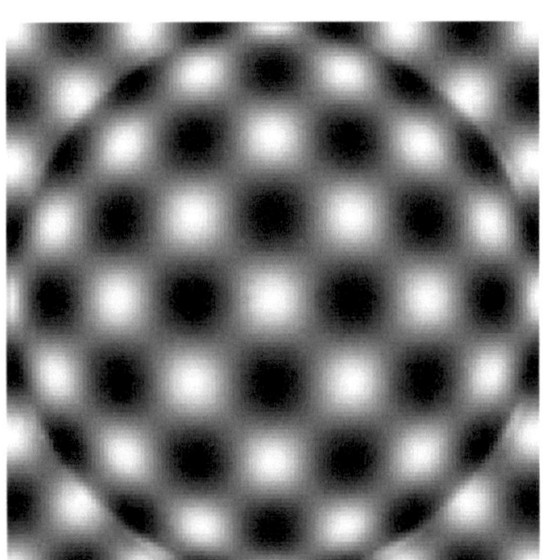

(a)

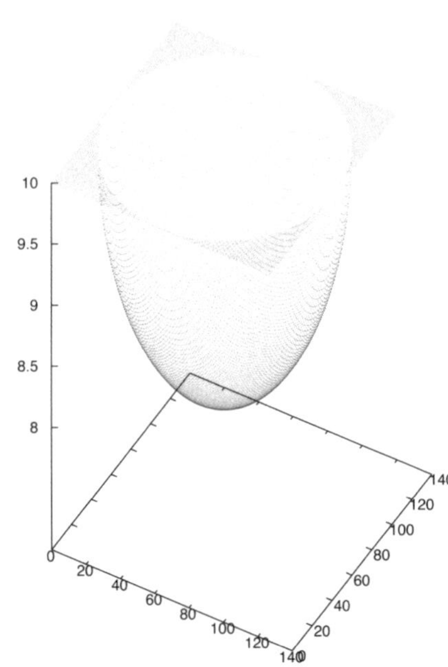

(b)

Figure 2.
Example of the data used in the experiments: (a) artificial image; (b) true depth map (reprinted from [13]).

them for depth recovery. The performance evaluation for f_t, f_x, and f_y actually measured from the image will be explained as the result of the real image experiment in Section 5.

We executed the proposed algorithm using $\sigma_r^2 = 10^{-4}$. Under this condition, the average size of the optical flow is about one pixel, which is sufficiently small compared to the size of the intensity pattern in the **Figure 2(a)**, which meets the conditions of the gradient method. A Gaussian random value with a zero mean and with a deviation corresponding to 1% of the mean of $\{f_t\}$ was added to the $\{f_t\}$ which completely satisfies Eq. (6) as observation noise. We evaluated the effectiveness of the smoothness constraint introduced by Eq. (8) by performing a depth recovery by varying the value of σ_d^2. The initial values of both σ_o^2 and σ_r^2 were 1.0×10^{-2} as arbitrary values, and $\{d\}$ was assumed initially as a plane of $Z = 9.0$. Examples of the results with $M = 100$ are shown in **Figure 3**. In addition, we varied M for each σ_d^2 to see the effectiveness of using many observations caused by small camera rotations together. The RMSE of the recovered depth for the values of σ_d^2 and M is organized in the **Table 1**. From these results, we can see that the smoothness constraint is important to reduce the degrees of freedom of $\boldsymbol{d}$. However, over-applying this constraint will increase the recovery error because the recovered depth map will be too smooth. Note that the scales of the Z axes in **Figure 3(a)-(c)** are different. We can also see that as the number of camera rotations increases, observation collection works well to improve recovery accuracy. In the future, it is desirable that σ_d^2 be estimated adaptively for each pixel or local region. We plan to treat it as a stochastic unknown variable and formulate it in the framework of variational Bayes scheme.

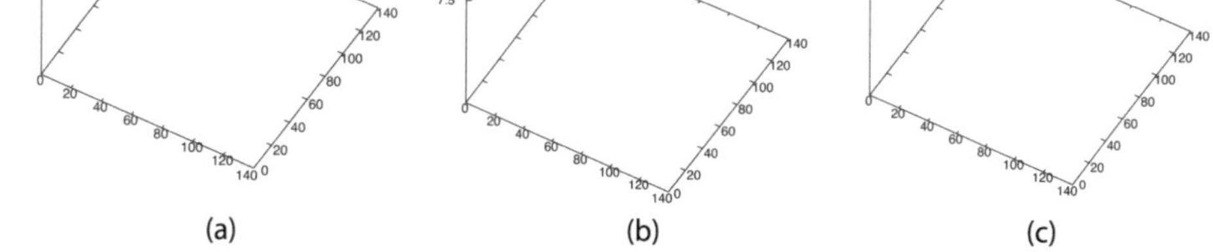

Figure 3.
Results of recovered depth maps with $M = 100$: σ_d^2 is (a) $\sigma_o^2 \times 10^{-1}$; (b) $\sigma_o^2 \times 10^{-3}$; (c) $\sigma_o^2 \times 10^{-5}$ (reprinted from [13]).

σ_d^2/σ_o^2	10^{-1}	10^{-2}	10^{-3}	10^{-4}	10^{-5}
$M = 50$	0.6983	0.3948	0.2097	0.0945	0.3699
$M = 100$	0.4741	0.3079	0.1841	0.0769	0.2124

Table 1.
RMSE of recovered depth.

4. Integral-type method

4.1 Image motion blurring related to depth

From Eqs. (2) and (3), and the probabilistic characteristics of $\boldsymbol{r}^{(j)}$, $\boldsymbol{v}$ is also a 2-D Gaussian random variable with $\mathrm{E}[\boldsymbol{v}] = \mathbf{0}$ and the variance–covariance matrix of

$$\boldsymbol{V}[\boldsymbol{v}] = \begin{bmatrix} x^2y^2 + (1+x^2+Z_0d)^2 & 2xy\left(1+\frac{x^2+y^2}{2}+Z_0d\right) \\ 2xy\left(1+\frac{x^2+y^2}{2}+Z_0d\right) & x^2y^2+(1+y^2+Z_0d)^2 \end{bmatrix}\sigma_r^2. \tag{19}$$

This equation can be seen as a function of the inverse depth d, so if we know the variance–covariance matrix at each pixel position, we can calculate the depth map.

There are some schemes to obtain the variance–covariance matrix of optical flow defined by Eq. (19) locally at each image position from multiple images observed through random camera rotations imitating tremor. The simplest and most natural way is to first detect the optical flow from the image and then calculate its quadratic statistic. However, here we are considering a case where the intensity pattern is fine with respect to the temporal sampling rate and it is difficult to accurately detect the optical flow. Therefore, we adopt an integral-formed scheme in which the variance–covariance matrix is calculated as the distribution of local image blur.

We define an averaged image brightness $f_{ave}(\boldsymbol{x})$ as an arithmetic average of observed M images $\left\{f_j(\boldsymbol{x})\right\}_{j=1,\cdots,M}$ with fixational eye movements. If M is asymptotically large, the following convolution holds using locally defined a two-dimensional Gaussian point spread functions $g_x(\cdot)$ and an original image brightness $f_0(\boldsymbol{x})$.

$$f_{ave}(\boldsymbol{x}) = \int_{\boldsymbol{x}'\in\mathcal{R}} g_x(\boldsymbol{x}')f_0(\boldsymbol{x}-\boldsymbol{x}')d\boldsymbol{x}', \tag{20}$$

where $\boldsymbol{x}$ indicates the image position, $\mathcal{R}$ is a local region around $\boldsymbol{x}$, and $g_x(\cdot)$ has a vairance-covaiance matrix indicated by Eq. (19). Furthermore, it is assumed that $\int g_x(\boldsymbol{x}')d\boldsymbol{x}' = 1$ is satisfied.

As explained above, we model $f_{ave}(\boldsymbol{x})$ as a blurred image due to fixation eye movements. The difference in the degree of blur of $f_{ave}(\boldsymbol{x})$ indicates the difference in depth. In other words, the closer the imaging target is to the camera, the greater this image blur.

4.2 Direct algorithms for integral-type method

In the two-step algorithm, after detecting the variance–covariance matrix of the image blur shown in Eq. (19), the maximum likelihood estimation of d using that as the observable is analytically possible [15]. However, the solution is not optimal because it observes indirect quantities. With optimality in mind, we need to use the direct method of directly estimating the depth map without determining $g\boldsymbol{x}(.;d)$. This strategy usually requires a numeric search or iterative update [16]. We constructed two algorithms as a direct method, each adopting local optimization and global optimization respectively.

A. Local optimization algorithm

We assume that a depth value in a local region $\mathcal{L}$ around each $\boldsymbol{x}$ is constant. Based on the minimum least square criterion, the objective function can be defined with respect to the depth corresponding to each pixel.

$$J_L(d(\boldsymbol{x})) \equiv \int_{\mathcal{L}} (f_m(\boldsymbol{x}-\boldsymbol{x}'') - f_{conv}(\boldsymbol{x}-\boldsymbol{x}''))^2 d\boldsymbol{x}'', \tag{21}$$

$$f_{conv}(\boldsymbol{x}) \equiv \int_{\mathcal{R}} g\boldsymbol{x}(\boldsymbol{x}'; d) f_0(\boldsymbol{x}-\boldsymbol{x}') d\boldsymbol{x}'. \tag{22}$$

We can recover the depth individually for each pixel by minimizing this function with respect to the depth corresponding to each pixel. Therefore, simultaneous multivariate optimization is not required and one-dimensional numerical search can be adopted.

B. Global optimization algorithm

By assuming a spatially smooth depth map in the solution, we can define the following objective function based on regularization theory of Poggio et al. [21].

$$J_G(d(\boldsymbol{x})) = (1-\lambda)\int (f_m(\boldsymbol{x}) - f_{conv}(\boldsymbol{x}))^2 d\boldsymbol{x} + \lambda \int \left\{ \left(\frac{\partial d(\boldsymbol{x})}{\partial x}\right)^2 + \left(\frac{\partial d(\boldsymbol{x})}{\partial y}\right)^2 \right\} d\boldsymbol{x}, \tag{23}$$

where λ is a parameter for adjusting the degree of constraint on the smoothness of the depth map, and the integration of Eq. (23) is performed for the entire image. From the variational principle, the Euler–Lagrange equation for $d(\boldsymbol{x})$ is derived using $\nabla^2 \equiv \partial^2/\partial x^2 + \partial^2/\partial y^2$ as follows:

$$\nabla^2 d = -\frac{1-\lambda}{\lambda}(f_m - f_{conv})\frac{\partial f_{conv}}{\partial d}. \tag{24}$$

For discrete computation, we can approximate the smoothness constraint in Eq. (23) using (i,j) as a description of an image position.

$$\begin{aligned}\left(\frac{\partial d(\boldsymbol{x})}{\partial x}\right)^2 + \left(\frac{\partial d(\boldsymbol{x})}{\partial y}\right)^2 \approx & \frac{1}{5}\left\{ (d_{i+1,j} - d_{i,j})^2 + (d_{i,j+1} - d_{i,j})^2 \right\} \\ & + \frac{1}{20}\left\{ (d_{i+1,j-1} - d_{i,j})^2 + (d_{i+1,j+1} - d_{i,j})^2 \right\}.\end{aligned} \tag{25}$$

Using Eq. (25) and the discrete representation of Eq. (24), we can minimize Eq. (23) by the following iterative formulation with an iteration number of n.

$$d_{i,j}^{(n+1)} = \overline{d}_{i,j}^{(n)} + \frac{1-\lambda}{\lambda}\left(f_{m,i,j} - f_{conv}\left(d_{i,j}^{(n)}\right)\right)\frac{\partial f_{conv}\left(d_{i,j}^{(n)}\right)}{\partial d}, \tag{26}$$

$$\overline{d}_{i,j}^{(n)} = \frac{1}{5}\left(d_{i+1,j}^{(n)} + d_{i,j+1}^{(n)} + d_{i-1,j}^{(n)} + d_{i,j-1}^{(n)}\right) + \frac{1}{20}\left(d_{i+1,j+1}^{(n)} + d_{i+1,j-1}^{(n)} + d_{i-1,j-1}^{(n)} + d_{i-1,j+1}^{(n)}\right). \tag{27}$$

4.2.1 Numerical experiments of integral-type method

The proposed algorithm assumes that the definition of the motion blur image in Eq. (20) holds. To observe such ideal motion blur, it takes enough exposure time for imaging. Here, we use artificial images to examine the characteristics of the proposed algorithm with respect to the relationship between the size of image motion and the fineness of intensity texture.

We artificially generate motion blur images. First, a true depth map is set up, and a large number of images are generated by a computer graphics technique that randomly samples $\boldsymbol{r}$ according to the Gaussian distribution in Eq. (4). By averaging these images, we can artificially generate motion blur images. Input motion blur images averaged $10,000$ images to mimic analog motion blur, **Figure 4** shows an example of a reference (original) image and a true inverse depth map. The image size is 256×256 pixels, which is equivalent to $-0.5 \leq x, y \leq 0.5$ measured in focal length units.

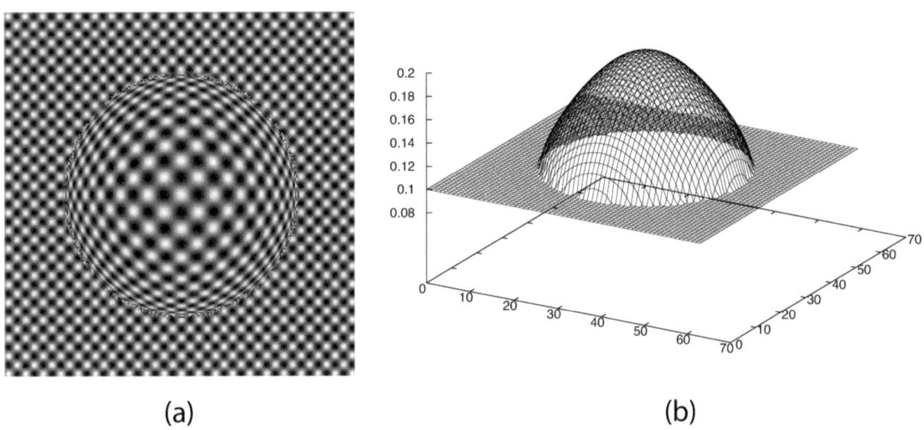

Figure 4.
Example of the artificial data used in the experiments: (a) original image; (b) true inverse depth map used for generating the blurred image (reprinted from [16]).

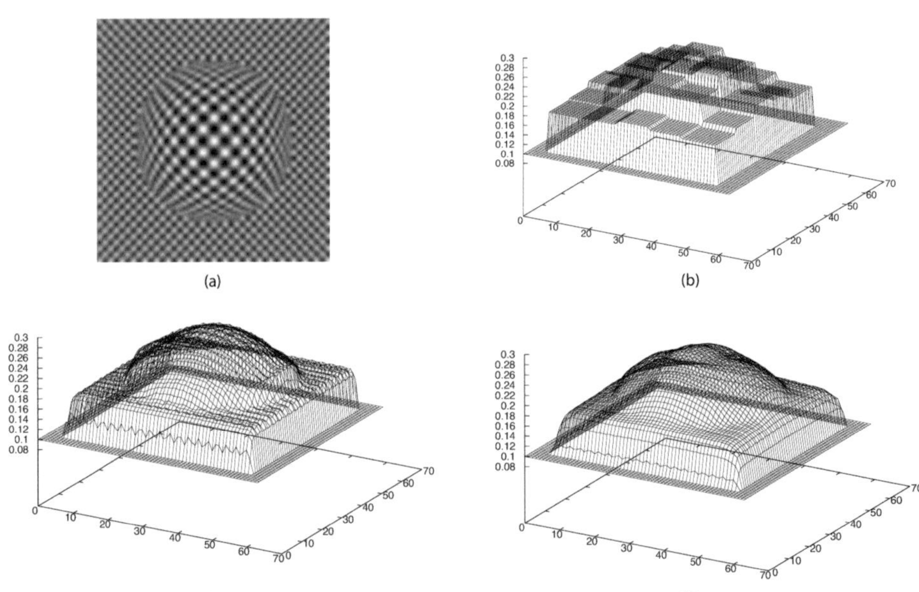

Figure 5.
Example of motion-blurred image and recovered inverse depth maps with $\sigma_r = 0.006$: (a) motion-blurred image; (b) local optimization; (c) $\lambda = 0.2$; (d) $\lambda = 0.6$ (reprinted from [16]).

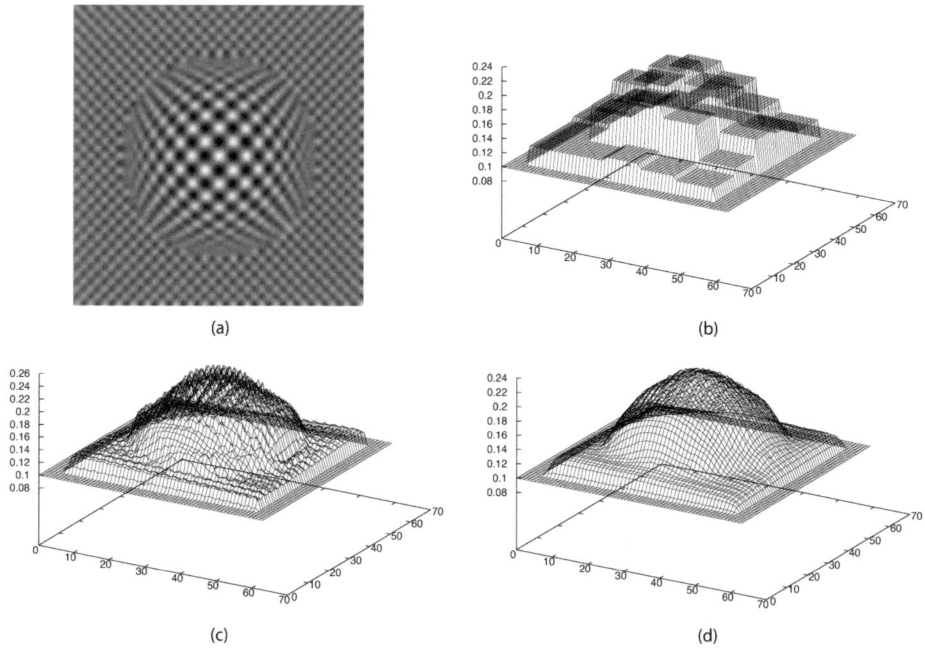

Figure 6.
Example of motion-blurred image and recovered inverse depth maps with $\sigma_r = 0.008$*: (a) motion-blurred image; (b) local optimization; (c)* $\lambda = 0.2$*; (d)* $\lambda = 0.6$ *(reprinted from [16]).*

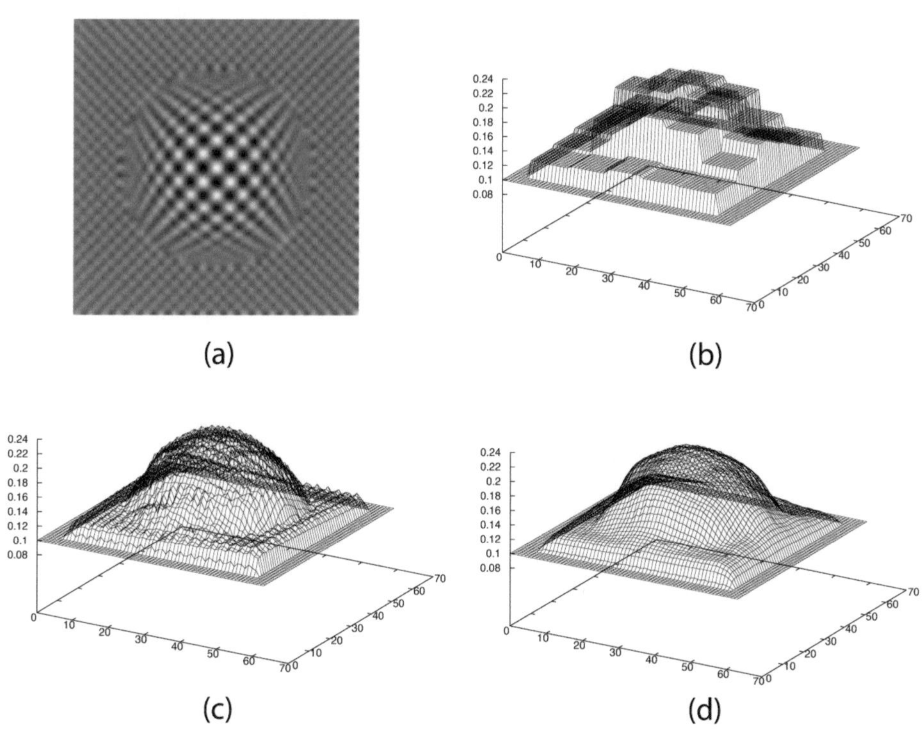

Figure 7.
Example of motion-blurred image and recovered inverse depth maps with $\sigma_r = 0.01$*: (a) motion-blurred image; (b) local optimization; (c)* $\lambda = 0.2$*; (d)* $\lambda = 0.6$ *(reprinted from [16]).*

Local optimization algorithms (LOA) are computationally expensive, and global optimization algorithms (GOA) are slow to converge. Therefore, we considered a hybrid algorithm that uses LOA sparsely to obtain the initial value of GOA. For the initial value of LOA, use the plane corresponding to the background of **Figure 4(b)**. To make a rough estimate, LOA uses blocks of 41×41 pixels as $\mathcal{L}$ in Eq. (21) and applies LOA once to each block. On the other hand, the size of $\mathcal{R}$ in Eq. (22) was adaptively determined according to the depth value updated by the optimization process. Therefore, $\mathcal{R}$ took a different size at each position in the image. We assumed a square area for $\mathcal{R}$. The length of that side was 10 times the larger of the two deviations of $g\boldsymbol{x}(\cdot; d)$ along x axis and y axis. These can be evaluated using Eq. (19).

The depth restoration simulation was executed while changing the camera rotation size σ_r. The recovered inverse depth map is shown in **Figures 5–7**. GOA uses various values of λ. The relationship between the root mean square error (RMSE) of the recovered depth map and the value of λ is also shown in **Figure 8**. From **Figure 5**, we can see that it is not suitable for depth recovery because it is difficult to accurately measure the motion blur of the image position with a small rotation of the camera. From **Figure 8(a)**, small rotations do not provide sufficient measurement information, so to reduce the RMSE of the recovered depth map, the smoothness constraint indicated by λ is strongly required. On the contrary, when the rotation is large, the point image distribution function becomes wider than the spatial change of the target shape. Therefore, the Gaussian function with the variance–covariance matrix in Eq. (19) is inappropriate, the motion blur recognized by this model is smoother than the actual blur of the image, and a depth recovery error occurs. This can be confirmed from the RMSE value in **Figure 8(c)**. Since the

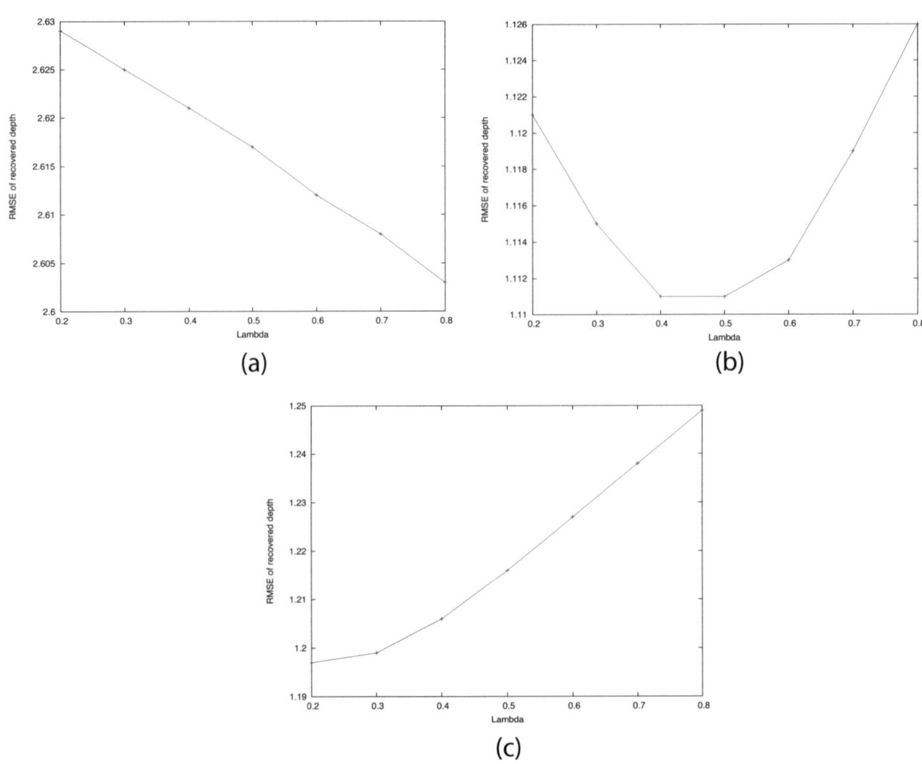

Figure 8.
Relation between RMSE of recovered depth and lambda: (a) $\sigma_r = 0.06$; (b) $\sigma_r = 0.08$; (c) $\sigma_r = 0.1$ (reprinted from [16]).

smooth depth tends to be recovered from the recognized smooth motion blur from **Figure 8(c)**, it can be confirmed that the smoothness constraint of Eq. (23) is an obstacle to the reduction of RMSE.

5. Real image experiments of differential-type method

5.1 Selective use of image pairs to improve accuracy

When applying the difference-type method to an actual image and checking the actual performance, the performance was improved by selecting the image pair used for depth restoration. We have adopted a scheme that excludes image pairs that are expected to have large approximation errors in the gradient equation on a pixel-by-pixel basis. We can use the inner product of the spatial gradient vectors of consecutive image pairs to select image pairs that do not cause aliasing problems. For each pixel, the image pairs of which the sign of the inner product $\boldsymbol{f}_s^{(i,j)\mathrm{T}}\boldsymbol{f}_s^{(i,j-1)}$ is negative are discarded. It is noted that $\boldsymbol{f}_s^{(i,j)} = \left[f_x^{(i,j)}, f_y^{(i,j)}\right]^{\mathrm{T}}$.

In the next step, from the image pairs remained by the above decision, we additively select the suitable image pairs at each pixel by estimating the amount of the higher order terms included in the observation of f_t. f_t is exactly represented as follows:

$$f_t = -f_x v_x - f_y v_y - \frac{1}{2}\left\{ f_{xx} v_x^2 + f_{yy} v_y^2 + 2 f_{xy} v_x v_y \right\} + \cdots. \tag{28}$$

After discarding a bad image pair, the higher-order terms can be considered small. In this case, the quadratic term in Eq. (28) can be estimated for each pixel i as follows:

$$-\frac{1}{2}\left\{\left(f_x^{(i,j)} - f_x^{(i,j-1)}\right)v_x^{(i,j)} + \left(f_y^{(i,j)} - f_y^{(i,j-1)}\right)v_y^{(i,j)}\right\}. \tag{29}$$

We can define a measure for estimating the equation error as the ratio of this higher order term to the first order term.

$$J = \frac{\left|\left(f_x^{(i,j)} - f_x^{(i,j-1)}\right)v_x^{(i,j)} + \left(f_y^{(i,j)} - f_y^{(i,j-1)}\right)v_y^{(i,j)}\right|}{2\left|f_x^{(i,j)} v_x^{(i,j)} + f_y^{(i,j)} v_y^{(i,j)}\right|}. \tag{30}$$

This measurement depends on the direction of the optical flow but is invariant with respect to the amplitude of the optical flow. To calculate the value of J, we need to know the true value of the optical flow. By examining the details of J, even if the difference of the spatial gradient $\boldsymbol{f}_s^{(i,j)} - \boldsymbol{f}_s^{(i,j-1)}$ is large, when the direction of $\boldsymbol{f}_s^{(i,j)} - \boldsymbol{f}_s^{(i,j-1)}$ is perpendicular to that of optical flow, the equation error becomes small. Therefore, the value $|\boldsymbol{f}_s^{(i,j)} - \boldsymbol{f}_s^{(i,j-1)}|/|\boldsymbol{f}_s^{(i,j)}|$ can be used as the worst value. In the following, the image pairs for which $|\boldsymbol{f}_s^{(i,j)} - \boldsymbol{f}_s^{(i,j-1)}|/|\boldsymbol{f}_s^{(i,j)}|$ is less than the certain threshold value are selected at each pixel to be used for depth recovery.

5.2 Camera system implementation

We built the camera hardware system for examining the practical performance of our camera model shown in **Figure 1**. The implemented camera system is shown in **Figure 9**.

Figure 9.
Camera system implemented for tremor rotations.

The camera system can be rotated around the horizontal axis i.e. X axis and around the vertical axis, i.e. Y axis. The rotation around the optical direction, i.e. Z direction, cannot be performed, which is not needed to gain the depth information. The parameters of the system are shown as follows: focal length is $2.8 - 5.0$ mm, image size is $1{,}200 \times 1{,}600$ pix., movable widths are 360 deg. for X axis and $(-10, +10)$ deg. for Y axis, and drivable minimum units are 1 pulse = 0.01 deg. for X-axis and 1 pulse = 0.00067 deg. for Y-axis.

5.3 Experimental results

In this section, we explain the results of the experiments using the real images captured by the developed camera system [22]. Our camera system has a parallel stereo function. That is, the camera can be moved laterally by the slide system. Prior to the experiment, we calibrated the camera's internal parameters, including focal length and Z_0, using the method in [23] and stereo calculations. The image used in the experiment is grayscale, consists of 256×256 pixels, and is 8-bit digitized. An example is shown in **Figure 10(a)**. The true inverse depth of the target object is shown in **Figure 10(b)**. It was measured in parallel stereo above using a two-plane model. In this figure, the horizontal axis shows the position in the image plane, and the vertical axis shows the inverse depth in units of focal length. We captured 100 images. The maximum number of iterations of the MAP-EM

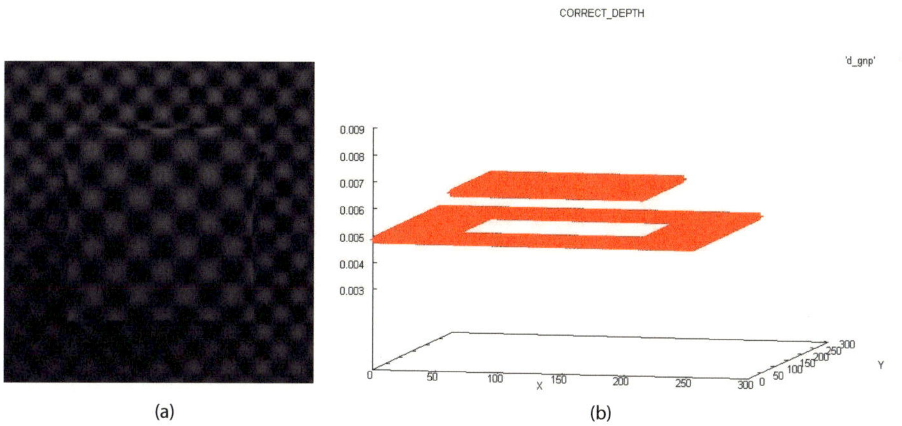

Figure 10.
Data for experiments: (a) example of captured image, (b) true inverse depth of object (reprinted from [22]).

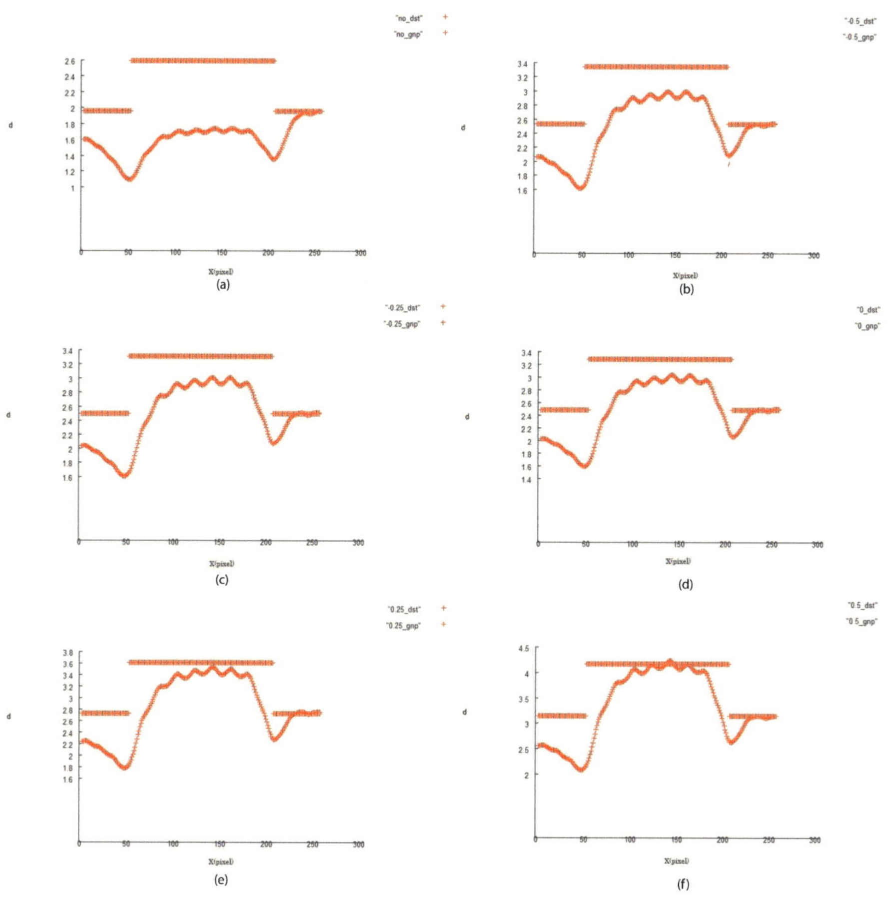

Figure 11.
Profiles of cross-section of recovered inverse depth: (a) all image pairs are used (100%), (b) threshold ×1.5(94% image pairs were used), (c) threshold ×1.25(86%), (d) threshold ×1(68%), (e) threshold ×0.75(62%), (f) threshold ×0.5(62%) (reprinted from [22]).

algorithm was set to 600. Within this number of iterations, the iterations of almost all experiments converged. σ_d^2 is heuristically determined. The average value of $|f_s^{(i,j)} - f_s^{(i,j-1)}|/|f_s^{(i,j)}|$ explained in the previous section with respect to all pixels was defined for each image pair as a standard magnification ($\times 1$) of the threshold for selecting the suitable image pairs. Namely, by decreasing the threshold magnification, we can discard more image pairs. Conversely, by increasing the magnification, many image pairs can be used for recovery. Because of the limit of pages, we only show the results with $\sigma_r^2 = 2.64 \times 10^{-2}$ by which the average of the optical flow's amplitude approximately coincides with $\lambda/4$.

Figure 11 shows the result of the recovered depth for each threshold set as a constant multiple of the reference value. We also looked at the results using all image pairs. From these results, it can be confirmed that by reducing the magnification, inappropriate image pairs can be discarded and the accuracy of depth recovery is improved. The percentage shown in the caption of the figure shows the number of image pairs used for recovery, which is determined in conjunction with the change in threshold.

6. Conclusions

In this chapter, we introduced a depth recovery algorithms that uses large number of images with small movements by using camera motion that simulates fixational eye movements, especially the tremor component. The algorithms can be divided into a differential-type and an integral-type. For the differential-type, it is desirable that the movement on the image is relatively small with respect to the texture pattern of the surface to be imaged, and conversely, for the integral-type, it is appropriate to apply it to a fine texture compared to the movement on the image. Therefore, ideally, the development of a depth recovery system in which both schemes function adaptively and selectively according to the target texture is the most important task in the future.

A detailed technical issue is to automatically determine the parameters that control the smoothness of the depth. This can be achieved by considering all unknowns as stochastic variables and formulating them in the variational Bayesian framework. As for the integration method, since the resolution of the recovered depth is low in principle, it is possible to consider a composite type in which the differential-type is applied again and refinement is performed on the result obtained by the integral-type.

So far, we have considered a method that assumes only tremor, but in the future, we are planning to study camera motion that also simulates drift and microsaccade. In the method for drift component, it is necessary to extend the method based on tremor to the online version, and then update the depth estimate while advancing the tracking of the target as time series processing. When using microsaccades, it is necessary to handle large movements between frames. Therefore, based on the correspondence of feature points, sparse but highly accurate depth restoration can be expected. Drift itself does not have much merit in its use, but it plays an important role in generating microsaccades. As described above, we believe that an interesting system can be realized by comprehensively using the three components.

On the other hand, stereoscopic vision and motion stereoscopic vision are difficult to handle objects with few textures. In [24], we proposed a stereo system that considers shading information. The projected images to both cameras are calculated by computer graphics technique while changing the depth estimation value. The depth is determined so that the generated image matches the image observed by each camera. As a result, the association between images is indirectly realized. By introducing this method, it becomes possible to handle textureless objects. We aim to develop a comprehensive depth restoration method, including the multi-resolution processing proposed in [12]. In another scheme that deals with the textureless region in stereo vision, the region where the depth value is constant or changes smoothly, called the support region, is adaptively determined [25]. We will also consider whether the relationship between image changes due to tremor and microsaccade can be used for adaptive determination of this support region.

In recent years, many realizations of stereoscopic vision and motion stereoscopic vision by deep learning have been reported [26–28]. And the relationship with the conventional method based on mathematical formulas is often questioned. The deep learning method is hampered by the addition of a large number of images and annotations to them. Although unsupervised learning is often devised, the solution is often limited. Therefore, even if the conventional method is rather complicated and takes time, if a method capable of more precise depth recovery is constructed, it can be used for annotation calculation of deep learning. This can be understood as copying the conventional method to deep neural network (DNN). DNN takes time to learn, but has the advantage of being able to infer at high speed. In this way, it is important that both schemes develop in a two-sided relationship.

Appendix

Here, the method of calibrating the axis of rotation is explained using **Figure 12**. Let a point in 3-D space be $\boldsymbol{X}_1 = [X_1, Y_1, Z_1]^{\mathrm{T}}$ in the coordinate system before camera rotation and $\boldsymbol{X}_2 = X_2, Y_2, Z_2]^{\mathrm{T}}$ in the coordinate system after rotation, and the coordinates of the corresponding points on the image be $\boldsymbol{x}_1 = [x_1, y_1, z_1]^{\mathrm{T}}$ and $\boldsymbol{x}_2 = [x_2, y_2, z_2]^{\mathrm{T}}$, respectively. Similarly, the optical axes before and after rotation are $\boldsymbol{z}_1^1 = [0, 0, 1]^{\mathrm{T}}$ and $\boldsymbol{z}_2^2 = [0, 0, 1]^{\mathrm{T}}$, respectively. If the rotation is taken around the X-axis, the rotation matrix is given by the following equation.

$$\boldsymbol{R} = \begin{bmatrix} 1 & 0 & 0 \\ 0 & \cos\theta & -\sin\theta \\ 0 & \sin\theta & \cos\theta \end{bmatrix}. \tag{31}$$

The translation $\boldsymbol{T}$ of the lens center generated by this rotation is given by the following equation in the coordinate system before rotation.

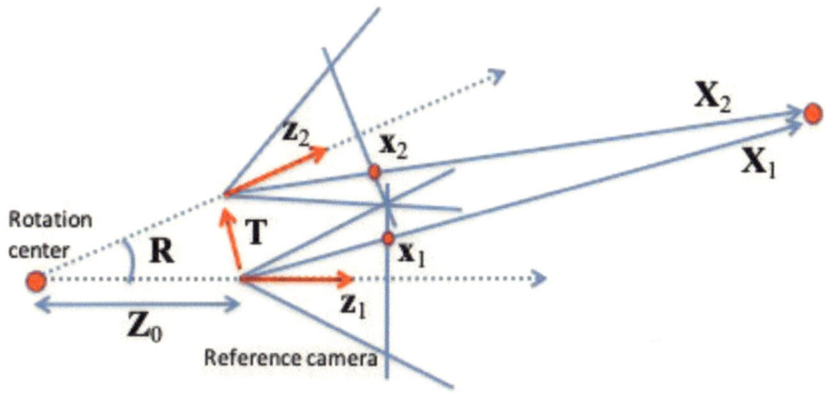

Figure 12.
Explanation of rotation axis calibration.

$$T^1 = Z_0 z_2^1 - Z_0 z_1^1 = Z_0(R - I) z_1^1 \equiv Z_0 S z_1^1, \tag{32}$$

where z_2^1 represents the optical axis after rotation in the coordinate system before rotation. In addition, X_1 and X_2 have the following relationship.

$$X_2 = R^{\mathrm{T}}(X_1 - T^1) \rightarrow RX_2 = X_1 - T^1 \tag{33}$$

Furthermore, by substituting the relation of $x_1 = X_1^1/Z_1$, $x_2 = X_2^2/Z_2$ into Eq. (33), the following equation is obtained.

$$Z_2 R x_2 = X_1 - Z_0 S z_1^1. \tag{34}$$

By expressing this equation in terms of components and organizing it, the following two equations are derived.

$$Z_2(y_2 \cos\theta - \sin\theta) = Y_1 + Z_0 \sin\theta, \tag{35}$$

$$Z_2(y_2 \sin\theta + \cos\theta) = Z_1 - Z_0(\cos\theta - 1). \tag{36}$$

By substituting Eq. (35) into Eq. (36), the solution of Z_0 can be derived as follows:

$$Z_0 = \frac{Z_1(y_2 \cos\theta - \sin\theta) - Y_1(y_2 \sin\theta + \cos\theta)}{\sin\theta(y_2 \sin\theta + \cos\theta) + (\cos\theta - 1)(y_2 \cos\theta - \sin\theta)}. \tag{37}$$

Author details

Norio Tagawa
Graduate School of Systems Design, Tokyo Metropolitan University, Tokyo, Japan

*Address all correspondence to: tagawa@tmu.ac.jp

References

[1] Martinez-Conde S, Macknik S L, and Hubel D. The role of fixational eye movements in visual perception. Nature Reviews. 2004;5:229–240.

[2] Stemmler M. A single spike suffices: the simplest form of stochastic resonance in model neuron. Network: Computations in Neural Systems. 1996; 61:687–716.

[3] Propokopowicz P and Cooper P. The dynamic retina. Int. J. Computer Vision. 1995;16:191–204.

[4] Hongler M-O, de Meneses Y L, Beyeler A, and Jacot J. The resonant retina: Exploiting vibration noise to optimally detect edges in an image. IEEE Trans. Pattern Anal. Machine Intell. 2003;25:1051–1062.

[5] Ando S, Ono N, and Kimachi A. Involuntary eye-movement vision based on three-phase correlation image sensor. In: Proceedings on19th Sensor Symposium; 2002. p. 83–86.

[6] Lazaros N, Sirakoulis G-C, and Gasteratos A. Review of stereo vision algorithm: from software to hardware. Int. J. Optomechatronics. 2008;5:435–462.

[7] Wang J and Zickler T. Local detection of stereo occlusion boundaries. In: Proceedings on CVPR; 2019. p. 3818–3827.

[8] Liu F, Zhou S, Wang Y, Hou G, Sun Z, and Tan T. Binocular light-field: imaging theory and occlusion-robust depth perception application. IEEE Trans. Image Process. 2019;29: 1628–1640.

[9] Horn B K P and Schunk B. Determining optical flow. Artif. Intell. 1981;17:185–203.

[10] Simoncelli E P. Bayesian multi-scale differential optical flow. In: Jahne B, Haubecker H, Geibier P, editors. Handbook of Computer Vision and Applications.Academic Press; 1999. vol. 2. p. 397–422.

[11] Bruhn A and Weickert J. Locas/kanade meets horn/schunk: combining local and global optic flow methods. Int. J. Computer Vision. 2005;61:211–231.

[12] Tagawa N, Kawaguchi J, Naganuma S, and Okubo K. Direct 3-d shape recovery from image sequence based on multi-scale bayesian network. In: Proceedings on ICPR; 2008. p. CD–ROM.

[13] Tagawa N. Depth Perception model based on fixational eye movements using Bayesian statistical inference. In: Proceedings on ICPR; 2010. p. 1662–1665.

[14] Tagawa N and Alexandrova T. Computational model of depth perception based on fixational eye movements. In: Proceedings on VISAPP; 2010, p. 328–333.

[15] Tagawa N, Iida Y, and Okubo K. Depth perception model exploiting blurring caused by random small camera motions. In: Proceedings on VISAPP; 2012, p. 329–334.

[16] Tagawa N, Koizumi S, and Okubo K. Direct depth recovery from motion blur caused by random camera rotations imitating fixational eye movements. In: Proceedings on VISAPP; 2013, p. 177–186.

[17] Sorel M and Flusser J. Space-variant restoration of images degraded by camera motion blur. IEEE Trans. Image Processing. 2008;17:105–116.

[18] Paramanand C and Rajagopalan A N. Depth from motion and optical blur with unscented Kalman filter. IEEE Trans. Image Processing. 2012;21:2798–2811.

[19] Dempster A-P, Laird N-M, and Rubin D-B. Maximum likelihood from incomplete data. J. Roy. Statist. Soc. B. 1977;39:1–38.

[20] Green P.-J. On use of the Em algorithm for penalized likelihood estimation. J. Roy. Statist. Soc. B. 1990; 52:443–452.

[21] Poggio T, Torre V, and Koch C. Computational vision and regularization theory. Nature. 1985;317:314–319.

[22] Tsukada S, Ho Y, Tagawa N, and Okubo K. Accuracy improvement for depth from small irregular camera motions and its performance evaluation. In: Proceedings on ICIAR; 2015, p. 306–315.

[23] Zhang Z. A flexible new technique for camera calibration. IEEE Trans. Pattern Anal. Machine Intell. 2000;22: 1330–1334.

[24] Wakabayashi K, Tagaw N, and Okubo K. Shape from multi-view images based on image generation consistency. In: Proceedings on VISAPP; 2013, p. 334–340.

[25] Wu W, Zhu H, Yu S, and Shi J. Stereo matching with fusing adaptive support weights. IEEE Access. 2019;7: 61960–61974.

[26] Laina I, Rupprecht C, Belagiannis V, Tombari F, and Navab N. Deeper depth prediction with fully convolutional residual networks. In: Proceedings on 3D Vision;2016, p. 239–248.

[27] Ummenhofer B, Zhau H, Uhrig J, Mayer N, Ilg E, Dosovitskiy A, and Brox T. DeMoN: Depth and motion network for learning monocular stereo. In: Proceedings on CVPR;2017, p. 5038–5047.

[28] Chen P-Y, Liu A H, Liu Y-C, Wang Y-C F. Towards scene understanding: unsupervised monocular depth estimation with semantic-aware representation. In: Proceedings on CVPR;2019, p. 2624–2632.

Chapter 2

Visual Identification of Inconsistency in Pattern

Nwagwu Honour Chika, Ukekwe Emmanuel, Ugwoke Celestine, Ndoumbe Dora and George Okereke

Abstract

The visual identification of inconsistencies in patterns is an area in computing that has been understudied. While pattern visualisation exposes the relationships among identified regularities, it is still very important to identify inconsistencies (irregularities) in identified patterns. The significance of identifying inconsistencies for example in the growth pattern of children of a particular age will enhance early intervention such as dietary modifications for stunted children. It is described in this chapter, the need to have a system that identifies inconsistencies in identified pattern of a dataset. Also, techniques that enable the visual identification of inconsistencies in patterns such as fault tolerance and colour coding are described. Two approaches are presented in this chapter for visualising inconsistencies in patterns namely; visualising inconsistencies in objects with many attribute values and visual comparison of an investigated dataset with a case control dataset. These approaches are associated with tools which were developed by the authors of this chapter: Firstly, ConTra which allows its users to mine and analyse the contradictions in attribute values whose data does not abide by the mutual exclusion rule of the dataset. Secondly, Datax which mines missing data; enables the visualisation of the missingness and the identification of the associated patterns. Finally, WellGrowth which explores Children's growth dataset by comparing an investigated dataset (data obtained from a Primary Health Centre) with a case control dataset (data from the website of World Health Organisation). Instances of inconsistencies as discovered in the explored datasets are discussed.

Keywords: missing data, contradictory, inconsistent, pattern, ConTra, visualisation, bad data

1. Introduction

It is often said that data is the lifeline of research. Due to the importance of data, several research areas such as machine learning, data science, data mining, data analytics and big data has been devoted to the full study and understanding of data. The use of data driven marketing (DDM) as an effective tool in determination of a strategic part of business management is proposed in [1] while the development of data-driven planning for management decision making is advocated in [2]. Also, there is a need for data driven research through open data source [3]. Also, it is noted in [4] that in order to effectively plan an experiment, there is need for preliminary data as a starting point. Even so, the need for valid data in research cannot be overemphasised. In fact, invalid and inconsistent data could inadvertently impart negatively on results of a research. The authors in [5] pointed out the importance of data validation for systematic software development. Similarly, the authors in [6] explained the importance of health records for diagnosis and treatment purposes. In general the need for valid data is indeed a concern that cuts across every research area. The study of big data has been found to have great impact on scientific discoveries and value creation [7]. The study continues to gain recognition as the quest for tools and measures for validating data continues. Also, [8] explains that the presence of noise hampers the induction of Machine Learning models from data, and can also make the training time longer. Noisy data according to [9] cannot be avoided but rather dealt with. Data, whether structured, semi-structured, or unstructured must be scrutinised with utmost care. The rigour of validating data could be tasking and are usually left in the hands of data scientists.

Data scientists acquire datasets from different environments which in most cases could be noisy. A noisy dataset contains uncertain and inconsistent data that could arise from missing values, imprecise data, and contradictory values, among others. The work of a data scientist includes among others, to explore big dataset in order to find interesting patterns and build supporting arguments for decision making. Such interesting patterns are likely to exclude noise in the form of conflicting or missing data in the dataset which do not support the arguments presented by the analyst. Data which are inconsistent with decision supporting facts should also be analysed. An approach to this analysis is to visually explore the patterns of the decision support data and associated inconsistencies.

Visual analytics is defined in [10] as the science of analytical reasoning facilitated by interactive visual interfaces. Data visualisation is useful for data cleaning, exploring data structure, detecting outliers and unusual groups, identifying trends and clusters, spotting local patterns among others [11]. The visual platform and representations enables better understanding and facilitates analytic as well as deductive reasoning. On the same note, visual analysis of data is important in understanding data and has been found to yield fruitful results in research. According to [12], visual analysis of data enables grasping the multidi-mensional "information reality" from the perspective of users. Visual analytics entails more than a mere visualisation. In fact, it can rather be seen as an integral approach to decision-making, combining visualisation, human factors and data analysis [13]. Visual analytics from another perspective is a data representation approach that employs interactive visualisation to integrate human judgement into algorithmic data-analysis processes [14]. Thus, visual representation of data plays a vital role in data interpretation and analysis.

It is important to analyse interesting patterns and associated noise from big datasets so as to identify the hidden patterns and knowledge in them. Unfortunately, some data scientists advocate deleting or not including the noisy data instead of visually depicting the noise and reporting the analyses. Certainly, deleting inconsistent data from a noisy dataset will increase the incompleteness in the dataset thereby reducing the soundness of the information retrieved from the dataset. Consequently, the noise in a dataset should be tolerated and its tolerance will enable the avoidance of losing interesting information about the dataset. The analysis of incomplete biological data of an organism for example, enhances the understanding of the abnormalities in the organism. Incomplete biological data existing in datasets from laboratory investigations such as data about genes and proteins provides clues to genetic disorder.

The importance of identifying inconsistencies in pattern can also be evident in survey dataset. A survey on pattern of menstruation can reveal a pattern that ladies between the ages of 20 to 30 years old who have not seen their menstruation for more than two months are pregnant. This pattern does not mean that all ladies of this same age bracket who have not seen their menstruations for more than two months from the survey data are pregnant. Obviously, there can exist ladies suffering from Hormonal aberrations for instance.

Also, respondents to survey questions may provide inaccurate responses, such as giving many consecutive items a response of "4" or repeating a pattern of "1, 2, 3, 4, 5...." as explained in [15]. Such purposefully deceptive or even contradictory responses are herein assessed as inconsistencies in patterns and should be portrayed visually as the wrong side of analysis. An example of inconsistency in a survey pattern involving giving many consecutive items a response of "4" is a pattern that shows responses that do not give many consecutive items a response of "4". It is therefore important to identify such inconsistencies in patterns of interest in order to properly provide analytical reports that expose the issues.

The importance of identifying and assessing inconsistent data is explained in works such as [15–17] but very few publications exist in the area of visually identifying inconsistencies in patterns of interest [18]. There is therefore a need to have a system that enables the visual analysis of inconsistencies in patterns of interest in a dataset. This is to provide data users with a holistic understanding of data of interest. It is stated in [18] "Of 612 data visualizations from 121 articles published online in February 2019 by a set of leading purveyors of data journalism, social science surveys, and economic estimates, 449 (73%) presented data intended for inference, but only 14 (3%) portrayed uncertainty visually, either by depicting explicit quantifications like intervals or conveying variance through raw data".

Consequently, the authors of this chapter emphasise the need for visualising inconsistencies in identified data patterns by explaining existing approaches and implementing novel approaches for visual analysis of inconsistencies in patterns. In Section 2, a detailed explanation on the concept of inconsistencies in pattern is given. In Section 3, two approaches for visualising inconsistencies in patterns are presented. The visual analyses of inconsistencies in objects with many attribute values and the visual comparison of an investigated dataset with a case control dataset is described. These approaches and their associated tools which were developed by the authors are discussed in the same section. The WellGrowth application is discussed in the same section. The WellGrowth app integrates the use of fault tolerance and colour coding to visualise inconsistent pattern while using data curated from Nsukka Medical Centre (NMC) and data from the website of World Health Organisation (WHO) as their control studies. A comparison of ConTra, Datax and WellGrowth Apps is presented in Section 4 while Section 5 is the conclusion and research focus for future work.

2. Inconsistencies in patterns

Any inconsistent data associated to a pattern reduces the quality of findings presented by the analyst about the pattern. An assessment of such inconsistent data can increase the trustworthiness of the findings from the analyst. There are everyday instances of inconsistent data in identified patterns which are likely to mar the patterns. Meade and Craig in [15] explain how inconsistent data from careless respondents of students' survey can be identified among data patterns common among respondents of the survey. Patterns derived from survey data can be associated to contradictory or incomplete responses. Also, patterns discovered in biological investigations can be associated to inconsistent and incomplete data. A gene expression dataset whose columns includes gene name, tissue name, expression and experiment ID can contain inconsistent data in an identified pattern where many experiments are performed for a particular gene in a particular tissue. An expression can be detected, not detected, or not available. If one of the interesting patterns is that a gene "xxx" is "detected" in experiments on tissue "yyy" of an organism at a particular developmental stage, then inconsistency of the pattern from the dataset will exist where there are data that shows that the gene "xxx" is "not detected" in other experiments that investigates the tissue "yyy" of the same organism at the same developmental stage. Also, uncertainty about the presence of the gene "xxx" can exist in the dataset where the information about the presence of the gene in the experiment about the tissue "yyy" at the same developmental stage is missing. Such missing information can be denoted by "unavailable" or empty space, among others. Inconsistent data relating to gene expressions in tissues of different developmental stages are reported in [17, 19]. Finally, a Radiologist chest x-ray report can be used to detect aortic unfolding which is mostly associated with systemic hypertension. However, there are instances of aortic unfolding which are not associated with systemic hypertension. There are also, some instances of aortic unfolding which it is not known if they are associated with systemic hypertension. These instances are inconsistent in a pattern involving systemic hypertension as a cause of aortic unfolding.

2.1 Visual analysis of inconsistencies in patterns of dataset

Inconsistent data which are associated to patterns in a large dataset can be difficult to visualise. This is because they are not explicitly indicated in the dataset as inconsistent. For example, missing data can exist as "unavailable", "forthcoming", "-", "not existing", or even empty spaces. Contradictions on the other hand, differ from one dataset to another, depending on the semantic definition of the data in the dataset. Interestingly, there are dedicated Applications such as CUBIST [19], ConTra [20], and R Package VIM [21] which enables the visualisation of the amount or pattern of contradiction and missingness in a noisy dataset. Inconsistent data whose pattern involves mutually exclusive type of contradictions is depicted by ConTra. Nwagwu explains in [20] how the contradictory attribute values in the gene "TSPAN6" of the tissue "Pancreas" is detected by ConTra and visualised in a pie chart. ConTra applies colour coding on charts to enable the visualisation of inconsistencies in a large dataset. Also, ConTra enables the visualisation of the pattern of distribution of contradictions across the dataset. It is further discussed in Section 3.11.

R Package VIM is a good analytical tool that focuses on visual presentations and analysis of missingness. It is used in plotting the aggregates of missingness in variables of a Barplots. It also shows missing data in a matrix plot, Histogram, Spline plot, Parallel coordinate plots and in Maps [21]. It uses Barplot to show the number and distributions of missing values for a sub-sample of the EU-SILC data from Statistics. Notwithstanding VIM's comprehensive collection of visualisation methods for exploring missing data, its environment requires extensive training in R skills in order to access its visualisation methods. Also, the VIM package does not enable the analysis of other types of inconsistencies such as contradictions in a dataset apart from the missingness.

There are other tools which enables the visualisation of inconsistencies as explained in [19, 22, 23]. A graphical tool is proposed in [22] that highlight inconsistent instances in the network such as the highlights of direct comparisons that strongly drive other treatment effect estimates and hot spots of network inconsistency. It also proposed a clustering approach that automatically groups comparisons for highlighting hot spots. CUBISTs [19] is an example of an application that applies colour coding and fault tolerance in traditional visualisation tools such as pie or bar chart to enable easy visual analysis of inconsistencies. Even so, these applications are not holistic in exploring inconsistencies in patterns and most of them are designed for particular domain of data analysis.

The analysis of inconsistencies in patterns of a dataset can be enhanced by adapting computational techniques such as fault tolerance and colour coding in traditional visualisation tools such as graphs to enhance the visualisation of inconsistencies in patterns. Fault tolerance necessitates the introduction of softness (statistical defined tolerance) in retrieving the inconsistencies in a dataset. Colour coding necessitates identifying the different ranges of inconsistencies with different colours. Section 3.0 presents how these computational techniques are used in computing inconsistencies in pattern as integrated in the approaches presented in this chapter.

3. Our approaches

Two approaches are presented for visualising inconsistencies in patterns in this section namely; visualising inconsistencies in objects with many attribute values and Visual comparison of an investigated dataset with a case control dataset. These approaches and their associated tools which were developed by the authors are discussed in this section.

3.1 Visualising inconsistencies in objects with many attribute values pattern

A dataset contains data about real world objects. These data contains objects which are associated to attributes and the attributes can be associated to single or many values. Real world objects 'G' such as house, book, car, and television are associated with different attributes 'M' which may have many values 'V'. A book (object) for example, can have colour (attribute) which can be black, white or brown (values). It can be established that particular object ($g \in G$) is associated with an attribute ($m \in M$) which contains many values. For example, a name (object) has marital status (attribute) such as married or single (values). Contradictory data can exist in a dataset when there is conflicting information such that an object ($g \in G$) that is associated with an attribute ($m \in M$), contains contradictory values such that m is associated with A and $\neg$A. An experiment (object) for example, can be associated with outcome (attribute) such as neutral, high, or low (values). A student (object) took a course (attribute) whose values can be absent, pass or fail. Some of the many valued attribute are likely to be mutually exclusive and should conform to mutual exclusion rule. The mutual exclusion rule can simply be stated that real world objects whose attribute values are mutually exclusive (meaning more than one attribute values cannot be associated with the object at once) are contradictory. Also, any attribute which do not contain the expected values is said to contain missing data.

Two open source tools are presented in this chapter to explain how to visualise inconsistencies in objects with many attribute values pattern namely ConTra and Datax. ConTra is discussed in an earlier publication [20] by some of the authors of this chapter and it is also discussed herein. Datax is another tool for highlighting inconsistency in patterns through mining and depicting missing data is presented in Section 3.12.

3.2 ConTra

ConTra[1] is an open source App developed by some of the authors of this chapter and it is used for mining contradictory data from attributes with many values

[1] https://github.com/ncjoes/contra

pattern where the contradictory data are objects associated with mutually exclusive attribute values. It enables its users to query attributes of particular objects whose attribute values are mutually exclusive and display the percentage of the values that contradicts the mutual exclusive rules and the percentage of the values that abide by the rule in a pie chart. Algorithm 1 displays the pseudocode for mining objects whose attribute values are contradictory and those whose attribute values are consistent.

ConTra was used to analyse objects in a Comma Separated Values (CSV) dataset containing over a million rows and six columns. The dataset 'Normal Tissue' dataset is deposited in [24] and it contains expression profiles for proteins in human tissues. It consists of the following columns: 'Gene', 'Gene name', 'Tissue', 'Cell type', 'Level', and Reliability. It has a size of 79.5 MB. Normal Tissue dataset reports experiments on tissues and identified gene expression levels such as low, medium, and high. It also indicates the annotated cell type ("Cell type") and the gene reliability. There can be multiple records for the same gene from different investigations (experiments) on the same tissues in Normal Tissue dataset.

Algorithm 1: ConTra's Algorithm for mining contradictory and consistent data as evident in [11]

1. Given a set of records in CSV format
2. Let G = Set of Objects from a selected column
3. Let M = Set of Attributes (titles of every column excluding the Object column)
4. Let O(a,b) = empty list where a = contradictory object index and b = contradictory attribute values
5. Let C(c,d) = empty list where c = consistent object index and d = consistent attribute values
6. For each Object 'g' in the set of objects 'G' which are associated to a set of mutually exclusive attributes 'M'
   i. If 'M' contains more than one mutually exclusive value then
      Store 'g' and also store each of the contradictory values in the list O(a,b)
   ii. Else
      Store 'g' in set of consistent objects and also store each of the consistent values in the list C(c,d)

End

7. Print contradictory objects O(a, b) and consistent objects C(c, d)

3.2.1 Evaluation of ConTra

ConTra was used to analyse the Normal Tissue dataset. Any experiment on a tissue in Normal Tissue dataset which indicates that its identified gene expresses more than one level of expression such as not detected, medium, high or low is inconsistent. As identified through the use of ConTra and discussed in [20], contradictions exist in two of the records (9.09%) of the gene 'TSPAN6' expression levels in the tissue 'Pancreas' in Normal Tissue dataset. This is depicted graphically in **Figure 1** as adopted from [20]. Evidently from **Figure 1**, it will be wrong to state that the pattern of expression of the gene 'TSPAN6' in the tissue 'Pancreas' of Normal Tissue dataset is of a particular level. This is because there are cases of contradictory expression in the associated data (TSPAN6 expression levels). Consequently, a holistic analysis of the expression levels of TSPAN6 on Pancreas in the Normal Tissue dataset should depict the existence of the contradictory data as shown in **Figure 1**.

ConTra provides a platform for visualising such inconsistencies in datasets whose objects exhibit a many attribute value pattern and are associated with mutual exclusive attribute values.

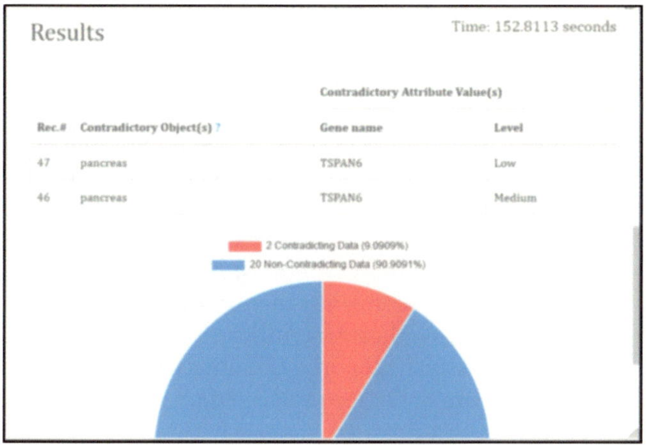

Figure 1.
Result of the analysis of the normal tissue dataset by ConTra's multiple attribute values approach.

3.3 Datax

Datax[2] is an open source application that mines missing data and associated patterns from a Comma Separated Values (CSV) Dataset. It is designed to enable the visualisation of the missing data in attribute values of a dataset by generating charts which depicts the incompleteness and any associated pattern. It has the following features:

- Ability to load and store CSV datasets for further visualisation
- Ability to display the statistics of incomplete data in a stored dataset
- Ability to visualise through matrix or bar plot, amount and distribution (patterns) of missingness in a selected dataset

The user of Datax can select attribute(s) or column(s) of interest from a dataset to visualise the missingness in them. Bar charts are programmed to use white lines to dynamically indicate the missingness in a dataset. Other important parameters measured in Datax include the number of columns in the investigated dataset and the percentage of missingness in each column.

Datax was used to mine incomplete data in an Amazon open source dataset[3]. The Amazon dataset has a size of 365.82 MB. It contains a list of over 1,500 consumer reviews of Amazon products such as the Kindle, and Fire TV Stick as provided by Datafiniti's Product Database[4]. It has a total of 27 columns which includes basic product information such as rating, review text, and more for each product. It also has a total of 1598 rows.

3.3.1 Evaluation of Datax

Datax was used to analyse the Amazon product review dataset as provided by Datafiniti's Product Database. **Figure 2** depicts the evaluation pane and shows a sneak peek into the first five rows of the investigated dataset while the right side

[2] https://github.com/marioJoker/Datax
[3] https://data.world/datafiniti/consumer-reviews-of-amazon-products
[4] https://datafiniti.co/products/product-data/

displays the statistical analysis of the dataset. The statistical analysis displayed includes: the names of the investigated columns, the total number of missing values per column of the dataset, and the percentage of the missing data per column (**Figure 3**).

As evident in **Figure 4**, the amount of missingness is indicated by the heights of the bars. Bars with equal height indicate joint missingness in investigated attributes. A Datax Bar plot reveals the amount of missing data and commonalities of such instances across the dataset. It can be observed in **Figure 4** that the columns V, W and Y have the same amount of missingness. Also the column headings U and X have the same amount of missing instances. Obviously, any analysis that includes columns whose pattern indicates significant amount of missingness should acknowledge such missingness in its reports. Columns that do not have missing data are also revealed in **Figure 4**. For example, columns A, B, C, D, F, G, J, M, N,

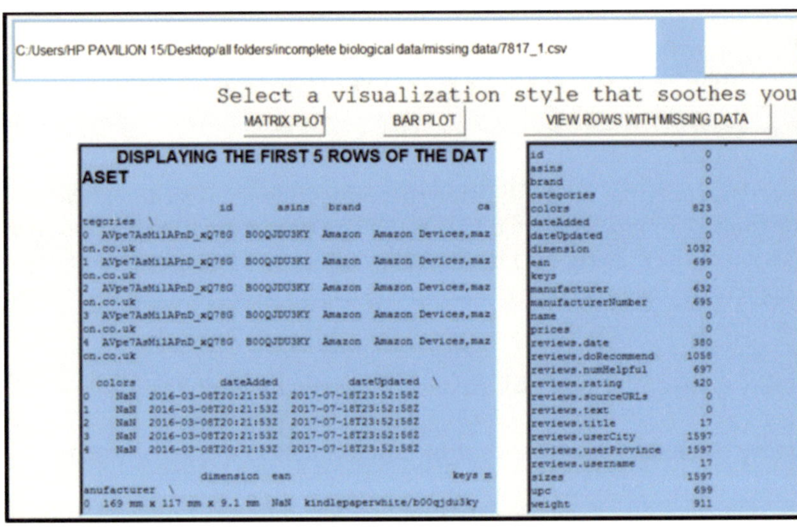

Figure 2.
Datax evaluation pane depicting the number of missing data per investigated column.

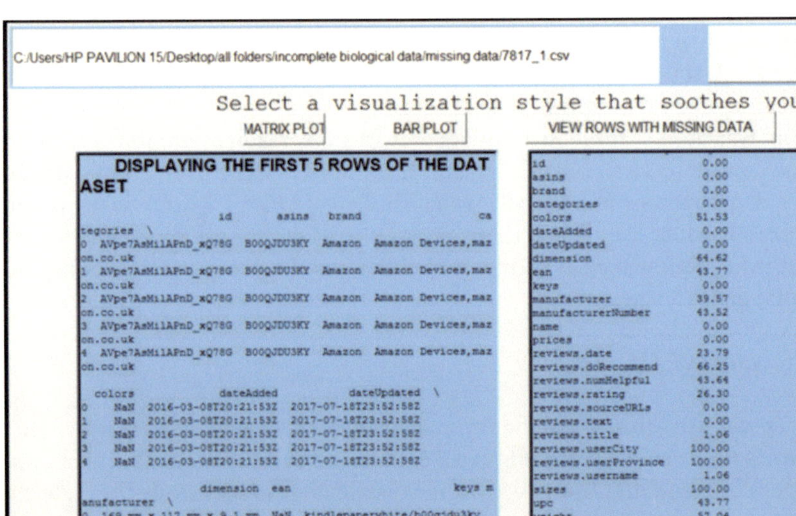

Figure 3.
Datax evaluation pane depicting the percentage of Missingness per investigated column.

S, and T do not have any missingness associated to them. Any assertion made by a data analyst about any column should first be evaluated for the relevance of missing data. Datax's Bar plot do not however, show the distribution of missingness among its investigated columns. This is explored by the matrix plot as depicted in **Figure 5**.

A Matrix plot of missing data as evident in **Figure 5** reveals the amount and distribution of missingness in the dataset. White colour is used for missing values while black colour is used for the available data values. It can be observed from **Figure 5** that the columns V, W and Y have 1597 missing data in common. The column headings U and X have equal amount of missing instances implying that each

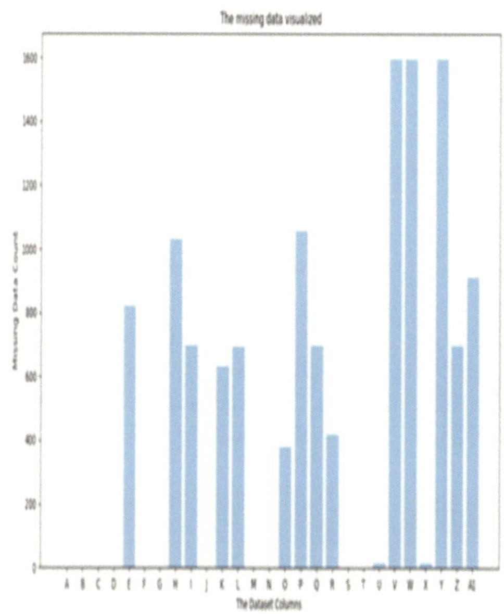

Figure 4.
Datax Bar plot depicting the amount of Missingness per investigated column.

Figure 5.
DataxMatrix plot depicting the amount and distribution of Missingness and available data per column.

reviewer that did not fill data in U, did not also fill data in X. The same observation holds for columns I and Z which have same distribution of missingness. The data analyst should make efforts to understand the relationships among the columns with joint and same distribution of missingness to present a robust report about the missingness in any discovered pattern.

Datax has also been used to evaluate cell phone reviews on the amazon online shopping store. The dataset is also deposited along Datax open source code[5]. It contains 11 columns and 1,048,576 records. Datax was evaluated by a team of software developers in University of Nigeria, Nsukka and they described its efficiency in mining missing data and visualisation of associated patterns as excellent. Even so, it does not visualise the different forms of missing data. It specifically mines empty cells without noting representations such as "-", "not existing", "not available", among others as missing data. The authors hope to integrate this ability in the next update of the application.

3.4 Visual comparison of an investigated dataset with a case control dataset

The visualisation of inconsistent data can be achieved through direct comparison of an investigating dataset with a case control dataset. Investigations that involve a comparison of an investigating dataset with a standard dataset are scenarios in which this approach can be used. This section of this chapter describes how WellGrowth app is used to enable the visual comparison of an investigated dataset with a case control dataset. It also describes the datasets investigated and how WellGrowth App was used in the investigation of the datasets.

3.5 Case control method

The case–control studies approach was used in comparing two datasets where one of the datasets is the case control while the other is the investigated dataset. World Health Organisation[6] (WHO) is the case control dataset and the dataset generated from Nsukka Medical Centre (NMC) is the investigated dataset. WHO data is gotten from children's empirical data which includes the length/height and weight of children at different stages of their growth for a sex matched reference. The weight and length of the children's data from WHO child growth standards for 0–12 months were used in investigating the NMC data. The average (50th percentile) score of the different children's weights and lengths in each month was used in the case control studies. This dataset is stored in WellGrowth app open source (see Section 3.22) for further analysis. The researchers collected the data (length for age and weight for age percentiles for girls and boys) directly from WHO web site (https://www.who.int/toolkits/child-growth-standards/standards/length-height-for-age).

NMC data are the weight and length of the children's growth data from 0 to 12 months for a sex matched references which are collected from the medical center Nsukka. The average (50th percentile) score of the different children's weights and lengths in each month as curated from NMC was used in the case control studies as the investigated dataset. This dataset is stored in WellGrowth app open source (see Section 3.22) for further analysis. The data collected from the NMC were not classified. So, the researcher classified them into different files according to the sex and the growth parameter per months from 0 to 12 months. The data were collected from NMC from May to august of 2020. **Table 1** presents a record of the number

[5] https://github.com/marioJoker/Datax/tree/master/amazon-cell-phones-reviews

[6] https://www.who.int/toolkits/child-growth-standards/standards/length-height-for-age

Sex	Height/Length	Weight
Girl	451	451
Boy	497	497

Table 1.
Number of children data collected according to the sex for the length and weight.

Age/Sex	Girls		Boys	
	Length	Weight	Length	Weight
0	54.75	3.95	52.5	4.3
1	57.5	5.449	58.75	5.85
2	61.5	6.75	63.75	7.1
3	62.75	6.5	65	7.05
4	63.5	6.7	66	7.6
5	64	6.8	66.5	7.1
6	65	7.4	66.75	8.1
7	65.78	8	67.2	9.5
8	66.25	7.5	67.7	9.5
9	67.25	8.4	70	8.9
10	67.8	8.8	72	9.5
11	68	8.5	72.45	9
12	68.5	8.4	72.9	9.5

Table 2.
The 50th percentile of data collected from NMC.

child's data collected by the researchers in NMC while **Table 2** presents the 50th percentile of the data collected from NMC.

3.6 WellGrowth

WellGrowth[7] app enables the visualisation of inconsistent data through direct comparison of an investigating dataset with a case control dataset. This is achieved through the visual evaluation of inconsistencies in children's growth pattern using the dataset from World Health Organisation[8] (WHO) as the case control dataset and dataset generated from Nsukka Medical Centre (NMC). These datasets are stored in the WellGrowth App for further evaluation by the App users.

WellGrowth adopts the average (50th percentile) score of WHO's children growth data for each month from 0 to 12 months in building WHO's growth curve. WHO's children growth data are gotten from children's empirical data such as height and weight at different stages of their growth for a sex matched reference. WellGrowth also integrates children growth data collected from the NMC. The average (50th percentile) score of the different children's weights and lengths in each month as curated from NMC are used to build the NMC/local growth curve. Finally, the individual growth curve is generated from inputs of a child's monthly weight/length as keyed into the WellGrowth input form by the WellGrowth App

[7] https://github.com/dora-png/growth-of-child
[8] https://www.who.int/toolkits/child-growth-standards/standards

user. A visual comparison of growth patterns from WHO to the growth patterns from NMC is used to enable the analysis of inconsistencies in children's growth data from Nsukka Medical Centre. Also, input of a child's growth data is used by WellGrowth app to enable a visual comparison of growth patterns of a Child with WHO's growth curve.

The authors designed WellGrowth for plotting growth pattern graphs from WHO, NMC and user's input data. WellGrowth App adopts colour coding and fault tolerance to enable easier visualisation of inconsistencies in their investigated datasets. For example, the average growth data whose value is less than 2 units from WHO's data value are yellow; the average growth data whose values are greater or equal to 2 and less than 4 are filled with red; while those values are greater or equal to 4 are filled with a yellow colour. Further details of WellGrowth implementations are expected in another publication by the authors.

3.7 Evaluation of WellGrowth app

Figure 6 presents individual, WHO and local (NMC) growth graphs showing the growth pattern (weight) of children whose ages are 0 to 12 months. **Figure 7** present a print view of Individual, WHO and local (NMC) WellGrowth's graphs. The average growth data whose values are less than 2 units from WHO's data value are indicated by yellow line while the average growth data whose values are greater or equal to 2 are filled with red line. **Figure 7a** shows the weight and **Figure 7b** shows the length of children whose ages are 0 to 12 months. It is evident from **Figure 7a** that there is no instance of inconsistency given the tolerance level of less than 2 units from WHO's a data value. The individual

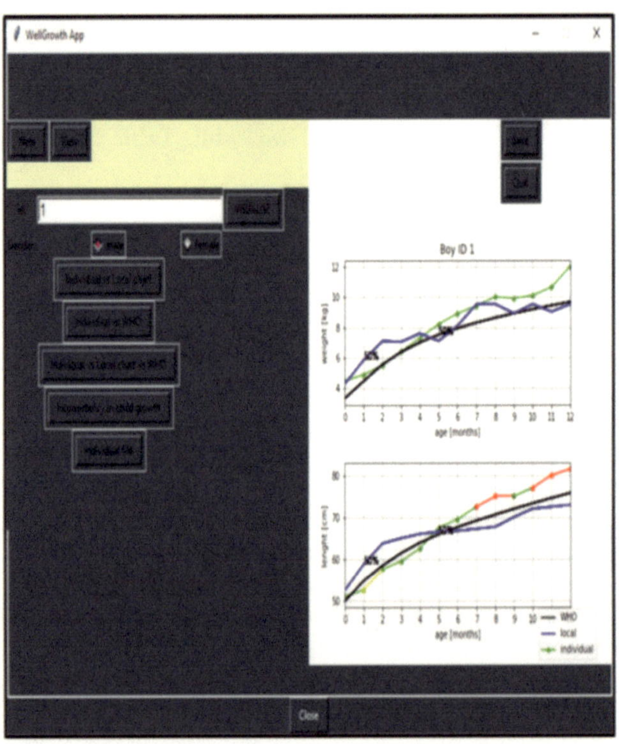

Figure 6.
WellGrowth app visualisation of growth patterns of individual vs. WHO vs. local (NMC) age graph after 12 months (weight in first graph vs. length in second graph).

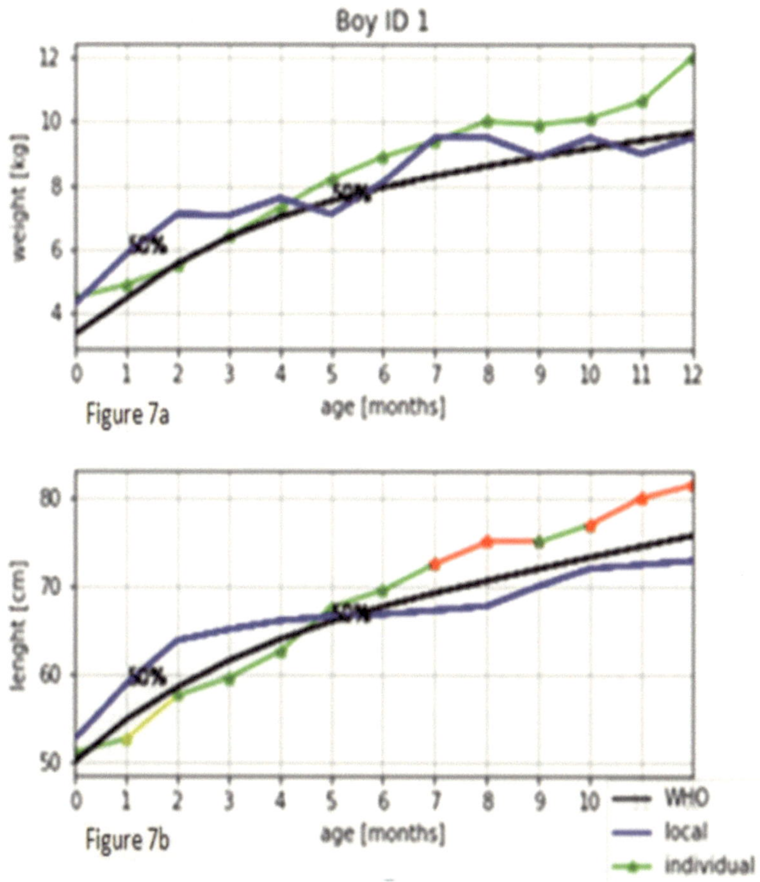

Figure 7.
*A print view of WellGrowth App's individual, WHO and local (NMC) growth graphs showing the weight (**Figure 7a**) and length (**Figure 7b**) growth patterns of children whose ages are 0 to 12 months.*

graph in **Figure 7b** is consistent with the WHO's growth pattern but there are issues of inconsistency in the 1st and 2nd month as indicated by the yellow line. Inconsistencies is depicted in **Figure 7b** individual line graph where the average growth data values are greater or equal to 2 and are indicated with red coloured line. The individual graph is consistent with the growth pattern of the WHO in the 5th to 7th month and 9th to 10th month but there are issues of inconsistencies in the 7th and 9th month and 10th to 12th month as indicated by the red line (see **Figure 7b**).

4. Comparison of ConTra, Datax and WellGrowth apps

Visual identification of inconsistencies in established patterns is achievable through data mining and analysis tools such as ConTra, Datax and WellGrowth apps. Each of these tools has its area of applicability depending on the kind of inconsistency explored. Datax for example, is most appropriately used for visualising patterns of missingness in CSV datasets unlike ConTra or WellGrowth that are used for mining and visualising contradictory data in patterns. **Table 3** presents a summary of the appropriateness of each of the tools in visualising inconsistencies in established patterns.

	ConTra	Datax	WellGrowth
Pattern of missingness	×	✓	×
Amount of missingness	×	✓	×
Amount of contradiction	✓	×	✓
Pattern of contradictory values	✓	×	✓
Colour coding	✓	×	✓
Fault tolerance	×	×	✓

Table 3.
Comparison of ConTra, Datax and WellGrowth apps.

Six yardsticks were used in comparing the appropriateness of the explored tools and they include: pattern of missingness, amount of missingness, amount of contradiction, pattern of contradictory values, colour coding, and fault tolerance. ConTra and WellGrowth for example, does not mine missingness nor explore the pattern of missingness in a dataset. They do not measure the amount of missingness, unlike Datax that is designed to evaluate both the pattern and amount of missingness using Matrix Plot and bar charts respectively. It is evident from our discussions in this chapter, that ConTra and WellGrowth apps are used to explore inconsistencies notably contradictory data in established patterns of interest. In doing this, WellGrowth apps adopt colour coding and fault tolerance while Datax only adopts colour coding. **Table 3** depicts these discussed yardsticks for comparing ConTra, Datax, and WellGrowth apps.

5. Conclusion and research focus for future work

This chapter has focused on the discussion of identifying inconsistencies associated with patterns. Even so, it has restricted its discussions to instances of contradictory data, deviations from standard data and missing values. Real life examples and open source datasets were used to illustrate our proposed approaches. The researchers anticipate that this interesting but understudied area of computing should be explored further by computer scientist to avoid instances of misinformation by our data analysts. Novel approaches for visual analysis of inconsistencies should be proposed. Also better means of diagrammatically visualising inconsistencies in pattern should be initiated.

Author details

Nwagwu Honour Chika[1*], Ukekwe Emmanuel[1], Ugwoke Celestine[2], Ndoumbe Dora[1] and George Okereke[1]

1 University of Nigeria, Nsukka, Enugu State, Nigeria

2 Nice diagnostic clinic Enugu State, Nigeria

*Address all correspondence to: honour.nwagwu@unn.edu.ng

References

[1] Veynberg RR, Timofeev A, Popov AA, Bortsova DE. Data driven marketing as a new approach to business development and sales methods. Espacios. 2018;39(12):3.

[2] Iyengar R, Mahal AR, Felicia UN, Aliyu B, Karim A. Federal policy to local level decision-making: Data driven education planning in Nigeria. International Education Journal: Comparative Perspectives. 2015;14(3):76-93.

[3] Krotoski AK. Data-driven research: open data opportunities for growing knowledge, and ethical issues that arise. Insights: the UKSG journal. 2012 Mar 7;25(1):28-32.

[4] Ezer D, Whitaker K. Point of View: Data science for the scientific life cycle. eLife. 2019 Mar 6;8:e43979.

[5] Patil MV, Yogi AN. Importance of data collection and validation for systematic software development process. Int'l Journal of Computer Science & Inf. Technology. 2011;3(2).

[6] Marinič M. The importance of health records. Health. 2015 May 5;7(05):617.

[7] Wang L, Alexander CA. Machine learning in big data. International Journal of Mathematical, Engineering and Management Sciences. 2016 Apr;1(2):52-61.

[8] Garćia LP, de Carvalho AC, Lorena AC. Noisy data set identification. InInternational Conference on Hybrid Artificial Intelligence Systems 2013 Sep 11 (pp. 629-638). Springer, Berlin, Heidelberg.

[9] Rao PS. Study and Analysis of Noise Effect on Big Data Analytics.

[10] Thomas JJ, Cook KA. A visual analytics agenda. IEEE computer graphics and applications. 2006 Jan 10;26(1):10-3.

[11] Unwin A. Why is data visualization important? What is important in data visualization?. 2.1. 2020 Jan 31;2(1).

[12] Cisek S, Krakowska M. Qualitative analysis of visual data in information behavior research. Zagadnienia Informacji Naukowej-Studia Informacyjne. 2019 May 6;57(1 (113)):7-25.

[13] Keim D, Andrienko G, Fekete JD, Görg C, Kohlhammer J, Melançon G. Visual analytics: Definition, process, and challenges. InInformation visualization 2008 (pp. 154-175). Springer, Berlin, Heidelberg.

[14] Cui W. Visual analytics: a comprehensive overview. IEEE Access. 2019 Jun 19;7:81555-73.

[15] Meade AW, Craig SB. Identifying careless responses in survey data. Psychological methods. 2012 Sep;17(3):437.

[16] Lin J, Keogh E, Lonardi S. Visualizing and discovering non-trivial patterns in large time series databases. Information visualization. 2005 Jun;4(2):61-82.

[17] Nwagwu HC, Orphanides C. Visual analysis of a large and noisy dataset. International Journal of Conceptual Structures and Smart Applications (IJCSSA). 2015 Jul 1;3(2):12-24.

[18] Hullman J. Why Authors Don't Visualize Uncertainty. IEEE transactions on visualization and computer graphics. 2019 Aug 19;26(1):130-9.

[19] Melo C, Aufaure MA, Orphanides C, Andrews S, McLeod K, Burger A. A conceptual approach to gene expression analysis enhanced by visual analytics.

InProceedings of the 28th Annual ACM Symposium on Applied Computing 2013 Mar 18 (pp. 1314-1319).

[20] Nwagwu HC, Okereke G, Nwobodo C. Mining and visualising contradictory data. Journal of Big Data. 2017 Dec;4(1):1-1.

[21] Templ M, Filzmoser P. Visualization of missing values using the R-package VIM. Reserach report cs-2008-1, Department of Statistics and Probability Therory, Vienna University of Technology. 2008 May 1.

[22] Krahn U, Binder H, König J. A graphical tool for locating inconsistency in network meta-analyses. BMC medical research methodology. 2013 Dec 1;13(1):35.

[23] White IR, Barrett JK, Jackson D, Higgins JP. Consistency and inconsistency in network meta-analysis: model estimation using multivariate meta-regression. Research synthesis methods. 2012 Jun;3(2):111-25.

[24] The Human Protein Atlas. http://www.proteinatlas.org/about/download Accessed 4 May 2020

Chapter 3

Build 3D Abstractions with Wireframes

Roi Santos Mateos, Xose M. Pardo and Xose R. Fdez-Vidal

Abstract

This chapter serves as an introduction to 3D representations of scenes or Structure From Motion (SfM) from straight line segments. Lines are frequently found in captures of man-made environments, and in nature are mixed with more organic shapes. The inclusion of straight lines in 3D representations provide structural information about the captured shapes and their limits, such as the intersection of planar structures. Line based SfM methods are not frequent in the literature due to the difficulty of detecting them reliably, their morphological changes under changes of perspective and the challenges inherent to finding correspondences of segments in images between the different views. Additionally, compared to points, lines add the dimensionalities carried by the line directions and lengths, which prevents the epipolar constraint to be valid along a straight line segment between two different views. This chapter introduces the geometrical relations which have to be exploited for SfM sketch or abstraction based on line segments, the optimization methods for its optimization, and how to compare the experimental results with Ground-Truth measurements.

Keywords: structure from motion, 3D abstraction, SLAM, 3D sketch, 3d reconstruction

1. Introduction

Most of the methods for environment abstraction from multiple views are just relying on points and ignoring other basic shapes like lines. Line based Structure from Motion methods based on lines create an abstraction based on straight line segments from a set of images. Analogously to point based abstraction methods like SIFT, in order to estimate the three-dimensional coordinates of lines in an spatial representation, the correspondences between lines among multiple images must be obtained by using detection and matching. The matching process for lines across the different views will return correspondences that can be exploited using 3D geometric relations. The matched features (points or lines) among views are used to estimate the position of cameras, referred to as extrinsic parameters. From the camera poses, the features are forward projected in the 3D abstraction or sketch.

The 3D line abstraction methods based on straight line segments that are most frequently found in the literature are designed to work altogether with detailed point-based reconstructions [1], therefore employing the camera extrinsic parameters obtained from these point rotation and translation invariants. This permits higher accuracy in the 3D reconstructions than using solely straight lines. A

 IntechOpen

different approach employs only straight line segment correspondences in the reconstruction, independently of point based 3D reconstructions [2]. This approach has been proved advantageous over the first one in scenarios where not enought feature points can be accurately put in correspondence between the different views. There are few publications about uncertainty analysis in 3D line reconstructions based on lines. One of the most recent ones explains the state of the art for these metrics [3].

Oppositely to points, straight lines have a direction, and this dimensionality can be exploited geometrically. The intersection of coplanar straight lines reveal geometrical information. Likewise, groups of segments will also indicate the location of the most probable vanishing points from a camera plane [4]. These geometric properties are not offered by points, therefore lines can be a good complement when performing a spatial reconstruction [5, 6]. Additionally, pairs of straight line segments are often related by the strong constraints of parallelism and orthogonality [7, 8]. This allows to combine individual similarities of pairs of segments altogether with the coplanarity constraints [9]. A recent work employed 3D lines to reconstruct surfaces [10].

1.1 SfM lines carry higher complexity

In literature, research about straight segments have always been developed after works related to feature points. Lines have often been left as a complement for applications of these works devoted to feature points. There are reasons for the line based SfM to be more complex than a feature point based one:

1. Detection of points is restrained to sole coordinates in images, while line detection extends to several pixels that are ideally adjacent to other pixels of the line. Nevertheless, in practice, detecting the limits of a straight line segment is not trivial in real images, due to digital noise, occlusions or changes in illumination. **Algorithms describing different continuity criteria must be employed in order to obtain a reliable edge detection in an image**. Moreover, as an straight line means a special case of an edge, detected edges have to be fit to straight lines. Fitting edges to straight segments can be accomplished by applying linear regression for the points comprising an edge in the image. Finally, the method has to find the endpoints of straight line segments, accounting for fragmentation or occlusions.

2. A set of pictures of the same scene may feature different kinds of viewpoint changes among captures, including camera rotations, zooms and translations. These **changes in the camera viewpoint produce a morphological transformation of the primitives in the captured frame**, which translates into displacements of the detected primitives, changes on their shape, distortions, fragmentation or even the impossibility to detect the same primitive in another image by employing the same operations that served to detect it in one of the pictures. Some of these transformations are not applicable to points, for instance a fragmentation: A point is either fully present or not, but it should not be such a thing as a detected fragmented point. Therefore, **there are more morphologic transformations that can affect 2D line segments than the ones that can affect points, due to camera viewpoint change**. Generally, prominent viewpoint transformations increase the difficulty in matching primitives, because the greater the transformation of the same primitives among different images of the scene, the greater the difficulty to match them.

3. Matching primitives between images is not always accurate, specially when dealing with line segments. When finding counterparts for primitives detected in other images, it is common to come out with several mismatched primitives. These wrongly matched primitives are referred to as **matching outliers** [11]. Matching outliers can produce that the description of the structure of groups of primitives can not be correctly compared to others, and employing inaccurate structure descriptions to propagate the matching to other images may cause problems when computing the final 3D abstraction. Some of the sources of the difficulties matching lines are because **line segments are subject to more morphological transformations than points. The description of individual lines are therefore more subject to these transformations, and less truthfully at the end. This fact forces line matching methods to rely more on structure of neighborhoods than points.** The accuracy of the description of these neighborhoods compared with the real morphological transformation of the lines it comprises are highly dependent on the ratio of matching outliers.

4. In the frame of 3D reconstruction from relations between feature points, known the relative position of two cameras and the position of one point on the first image, **there is a constraint that forces the counterpart of this point on the second image to lay on a line. It is called epipolar constraint. But a single infinite 2D line represented in two images does not feature epipolar constraint. The only point-to-point valid correspondences in matched segments under a viewpoint change are their endpoints.** For this case of a line segment, in order to estimate its position in 3D, is required to detect in the images both endpoints of the line segment. In some cases it may be difficult to accurately detect the end of a line segment in an image. For instance, a segment can end by merging with another edge under a different slope, progressively dimming until it vanishes, by intermittent occlusions, or being abruptly fragmented. Moreover, one or both segment endpoints may lay in the limits of the frame, and in this case it will not be possible to extract the 3D pose of the line.

The above mentioned tasks portrait the main differences between lines and points raised out during the engineering of a complete line-based 3D sketch generation method from images. For each stage of the method,specific tasks and problems have to be solved in the state-of-the art: detection of borders, matching lines over pairs of views, comparing the line matching performance against competition, relate the matched primitives among sets of more than two images, estimation of spatial lines, optimizing the abstraction and exploiting the resulting 3D structure.

2. Estimate 3D straight line segments

A 3D line can be thought as a multi-view entity that relates a perceivable line segment in the real world to its counterparts in images, given that these have been correctly detected and matched. The process of generating a 3D representation from different pictures of the scene is visually represented in **Figure 1**.

For the SfM problem, the poses of the cameras that took the pictures are not provided, and it is up to the SfM algorithm to simultaneously estimate the poses for the cameras and primitives. In the present case, SfM has to estimate the pose of the lines in space, relative to the cameras. The first requirement for the method is the calibration matrix K for each camera, which provides the transformation between

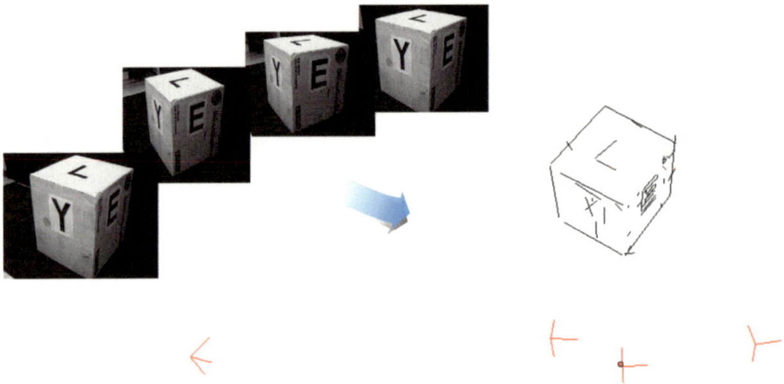

Figure 1.
Visual representation from [2]. It depicts the challenge of converting a set of 4 pictures into a 3D sketch featuring the line segments and camera axis. The 4 cameras are represented as three axis reference frame in red.

each point in one image, in homogeneous coordinates, to a ray in Euclidean three-dimensional space. Secondly, SfM has to estimate the projection matrices P for the cameras, representing a map from 3D to 2D:

$$\mathbf{x} = \mathbf{P}\mathbf{X}, \tag{1}$$

where $\mathbf{x}$ is a 2D point on the image, and $\mathbf{X}$ its projection in 3D space. $\mathbf{K}$ is intrisic to each camera, while $\mathbf{P}$ is extrinsic and embeds the 3D translation and rotation of the camera's image plane. The estimated translation is valid up to scale.

A common space can be built to host the cameras and spatial lines. For this new common space the camera that took the first processed picture takes the place of the origin, and for the rest of cameras $\mathbf{P}$ can be estimated from the lines matched between the captured images. Alternatively, camera poses can also be retrieved from a feature-point based SfM pipeline and these cameras be employed for the estimation of spatial lines. For instance, the feature-point descriptor SIFT [5] can be used to match points in images with a low ratio of outliers. These feature point relations are obtained both in the foreground and background. A set of relations between points or lines in two images allows to estimate the homography constraints between both views by applying the 5-point algorithm [12] using the points or the segment endpoints. A purge of outliers can be performed employing RANSAC [13] for robust estimation. Therefore, a set of stereo 3D projections is obtained combining the available images pairwisely, and each stereo system featuring both camera poses and a point cloud. The objective is to obtain an unique 3D point cloud sketch, embedding all cameras and point matches. Hence, camera poses are sequentially stacked, relative to each other, in the new spatial reference frame. And the 3D estimations for the feature points in the new 3D space can be computed as the center of gravity for their position relative to the common camera in both stereo systems. Finally a sparse bundle adjustment [14] is used to minimize the pixel distance of the back-projected 3D point and the original observation of this point on each image in homogeneous coordinates. These reprojection errors on the planes of the cameras are minimized employing the Levenberg-Marquardt algorithm. The resulting keypoint-based 3D reconstruction contains the optimized 3D estimations for the cameras and the point cloud.

Several straight segment matching methods are based on texture descriptors [15, 16], coloring [17] or in keypoint-line projective invariants [18, 19]. Under these conditions, matching results will be influenced by the level of texture in the images. In the case that a low number of detected segments can be distinguished by employing image texture based descriptor, or in case that a low number of feature points are identified throughout the set of images, the resulting set of matched lines will not be satisfactory. On the other hand, if line matching is rooted on weak epipolar constraints [1], line matching will be highly dependent on the accuracy of the camera poses.

Extrinsic parameters for cameras are needed to project the matched lines into space. Having the same segment completely detected and without fragmentation for both views under viewpoint change, endpoints are the only points in a segment with known exact counterpart in the other image. Unfortunately, segment detection is not accurate in the location of the endpoints. Therefore, the most accurate abstractions will be the ones built rooted on camera extrinsics obtained from a dense feature point based SfM. As written above, known the projection matrices P of two cameras, a point on an image projects as a 3D ray in Euclidean space. And this 3D ray projects like an infinite 2D line on any plane different than the one that contains the point. Therefore, each 3D point X_p will have its image into an epipolar line e_p contained in the image. As the unknown point is constrained into a line in the other image plane, analogously a segment will be constrained between both epipolar lines corresponding to the segment endpoints. This weak epipolar constraint can be employed for matching segments between images [1].

3. Geometric relations

A *3D abstraction method* estimates the position of 3D line segments $\Gamma = \{\Gamma_1, \Gamma_2, \Gamma_3, ..., \Gamma_N\}$, from an unordered sequence of images, taking from cameras with planes $\Upsilon = \{\Upsilon^1, \Upsilon^2, \Upsilon^3, ..., \Upsilon^M\}$. Straight lines are detected in the original images, put in correspondence among them, forward projected into space, and rewritten in homogeneous coordinates.

The 3D line based sketch $\{\Upsilon, \Gamma\}$ is built from the knowledge of correspondences among line projections l on camera planes, and the intrinsics of all the cameras. The following paragraphs explain the linear triangulation of these observations, as performed from scale-space images. This allows to discriminate and weight down lines that have been detected on two or more scales with a different slope. The practical consequence is that prior to any 3D extrapolation of the observed lines, matching inliers with inconsistent endpoint location among scales on both images can be avoided, as these lines might introduce uncertainty in the estimation for the pose of the camera.

The camera poses **P** are estimated from the endpoint correspondences of l. The Essential matrix **E** is computed from the camera pairs, by using the Five-Point Algorithm [12], and RANSAC [13] for hypotheses generation. Having **E** and l, the relative camera rotation and translation among the first pair of cameras $\boldsymbol{P}^j = [\mathbf{R}|\mathbf{t}]$ are estimated using cheirality check and discarding the triangulated endpoints that are not in front of the cameras. The left camera is chosen to have the pose $\boldsymbol{P}^1 = [\mathbf{I}|0]$, and the newly added cameras are stacked from this position in the unique reference frame.

The forward projection of lines in 3-space is described in the page 196 of Hartley and Zisserman's book [20]. The 3D forward projection $\mathbf{\Gamma}_\mathrm{i}$ of a line, bundled in the same reference frame, can be obtained using the *DLT* method on the set of stereo 3D camera back-projections. This is performed in homogeneous coordinates because it allows to consider line endpoints in the infinite. Therefore, from now on, when a 2D point is mentioned it will be supposed homogeneous coordinates. There exists a 3×3 matrix **E**, known as the essential matrix, such that if u and u' are a pair of matched points, then $u'\mathbf{E}u = 0$. If a sufficient number of matched points are known, the matrix $\mathbf{E}$ may be computed as the solution of an overdetermined set of linear equations. For the present problem, the internal calibration of the cameras is known, therefore it is possible to determine from $\mathbf{E}$ the relative placement of the cameras and hence the relative locations of the 3D points corresponding to the matched points. A linear triangulation method is projective-invariant because only camera and line distances are minimized.

The above described DLT method for lines starts with the segments on the pair of cameras $\{\Upsilon^a, \Upsilon^b\}$ with the highest inlier ratio. Based on this first triangulation, the other cameras are appended to the 3D abstraction: The next camera Υ^c is chosen according to the higher inlier ratio of line matching with Υ^a and Υ^b. Analogously, the following camera Υ^n is picked among the ones with the higher inlier ratio of line matching with previously selected cameras. The detection of 2D lines l in the original images carry an uncertainty for the position of these observations. This uncertainty implies that no 3D point $\boldsymbol{X}$ will satisfy that their projections on cameras Υ^1 and Υ^2 are $\boldsymbol{x_1} = \boldsymbol{P}^1\boldsymbol{X}$, $\boldsymbol{x_2} = \boldsymbol{P}^2\boldsymbol{X}$ respectively. Moreover, the image points do not satisfy the epipolar constraint $\boldsymbol{x_2 F x_1} = 0$. Therefore, a method that only minimizes the distances on the image from the estimations to the observations is required: A projective-invariant triangulation method. A linear triangulation [20] method does not depend on the projective frame in which $\boldsymbol{X}$ is defined.

The forward projection from a normalized 2D line observed on the camera image plane m, denoted by $\boldsymbol{l_i^m}$, is the plane $\boldsymbol{P_m}^T\boldsymbol{l_i^m}$, so the condition for a point $\boldsymbol{X}_a$ to be in this plane is:

$$\left(\boldsymbol{l_i^m}\right)^T \boldsymbol{P}_m \boldsymbol{X}_a = 0. \tag{2}$$

Each point $\boldsymbol{X}_a$ returns a linear equation in the entries of $\boldsymbol{P_m}$. Denoting by $\boldsymbol{x}^i_{m,E}$ and $\boldsymbol{x}^i_{m,F}$ the forward projection of the endpoints of $\boldsymbol{l}^i_m$, named $\boldsymbol{X}^i_E$ and $\boldsymbol{X}^i_F$, under $\boldsymbol{P}_m$, then any other 3D point on the line $\boldsymbol{X^i}(\mu) = \boldsymbol{X}^i_E + \mu\boldsymbol{X}^i_F$ projects to a point:

$$\boldsymbol{x}^m_i(\mu) = \boldsymbol{P}_m(\boldsymbol{X}_{i,E} + \mu\boldsymbol{X}_{i,F}) = \boldsymbol{x}^m_{i,E} + \mu\boldsymbol{x}^m_{i,F}, \tag{3}$$

which is on the line segment $\boldsymbol{l_i^m}$.

In the described method, an unique reference frame is built. The world reference system is fixed onto the first camera, hence its camera matrix, $\boldsymbol{P_E}$, is computed with $\boldsymbol{R_E} = \boldsymbol{I}$ and $\boldsymbol{T_E} = \boldsymbol{0}$. The extrinsics for the partner camera $\boldsymbol{P_m}$ on the baseline is obtained from the essential matrix by using RANSAC. Before the subsequent DLT triangulations with a new camera, its extrinsics are estimated also by RANSAC from the 2D-3D results of the already computed DLT. From here, new cameras will be added incrementally, just one per DLT iteration, in order to avoid DLTs between two uninitialized camera projection matrices.

For DLT it is required a set of observed line correspondences, $\boldsymbol{l}^m_j$ to $\boldsymbol{l}^n_j$, matched among images. The projection on the image plane of camera m of an endpoint $\boldsymbol{X}_{i,E}$ of the spatial line $\boldsymbol{\Gamma}_j$ is denoted as $\boldsymbol{x}^m_{j,E} = \boldsymbol{P}_m\boldsymbol{X}_{j,E}$. This point on the m-th camera plane is matched to its counterpart on the n-th camera $\boldsymbol{x}^n_{j,E} = \boldsymbol{P}_n\boldsymbol{X}_{j,E}$. Both equations can be combined into $\boldsymbol{AX}_{j,E} = 0$, where $\mathbf{A}$ is the matrix of equation coefficients. It is built from the matrix rows $\boldsymbol{A}_r$ contributed from each correspondence, whose resemble the movement of each line between both views. $\boldsymbol{X_{j,E}}$ contains the unknowns for the endpoint position.

By using the cross product on the m-th camera: $\boldsymbol{l}^m_j \times (\boldsymbol{P}_m\boldsymbol{X}_{j,E}) = 0$,

$$x_m\left(\boldsymbol{p}_m^{3T}\boldsymbol{X}_{j,E}\right)-\left(\boldsymbol{p}_m^{1T}\boldsymbol{X}_{j,E}\right)=0, \tag{4}$$

$$y_m\left(\boldsymbol{p}_m^{3T}\boldsymbol{X}_{j,E}\right)-\left(\boldsymbol{p}_m^{2T}\boldsymbol{X}_{j,E}\right)=0, \tag{5}$$

$$x_m\left(\boldsymbol{p}_m^{2T}\boldsymbol{X}_{j,E}\right)-y_m\left(\boldsymbol{p}^{1T}\boldsymbol{X}_{j,E}\right)=0 \tag{6}$$

where (x_m, y_m) and (x_n, y_n) are the coordinates of $\boldsymbol{x}^m_{j,E}$ and $\boldsymbol{x}^n_{j,E}$ respectively. $\boldsymbol{p}_m^{rT}$ is the r-th row of $\boldsymbol{P}_m$. It can be decomposed similarly for $\boldsymbol{P}_n$, and compose the equation of the form $\boldsymbol{AX}_{j,E}=0$. Solving:

$$\boldsymbol{A}=\begin{bmatrix} x_m\boldsymbol{p}_m^{3T}-\boldsymbol{p}_m^{1T} \\ y_m\boldsymbol{p}_m^{3T}-\boldsymbol{p}_m^{2T} \\ x_n\boldsymbol{p}_n^{3T}-\boldsymbol{p}_n^{1T} \\ y_n\boldsymbol{p}_n^{3T}-\boldsymbol{p}_n^{2T} \end{bmatrix}. \tag{7}$$

The solution for the 4 equations of the over-determined problem (four equations for four homogeneous variables) is only valid up to scale. The set of points in space mapping to a 3D line $\boldsymbol{\Gamma}_j$ via $\boldsymbol{P}_m$, is the plane $\boldsymbol{P}_m\boldsymbol{\Gamma}_j$.

The result of the linear triangulation process is $\boldsymbol{\Gamma}_\mathbf{i}$ and u^j, represented in cartesian coordinates.

Every 3D segment Γ_i is estimated as the center of gravity of the estimations for the same line for each par of images. The set of line projections observed in Υ is represented as $l = \{l_1^1, l_2^1, \ldots, l_N^1, \ldots, l_N^M\}$. A Line Feature is defined as a subgroup of projections from l of the same 3D line Γ_i. The Line Features are noted as $L = \{L_1, L_2, \ldots, L_N\}$. The 3D lines Γ are obtained by forward projecting the endpoints of l from pairs of camera planes of Υ, by using linear triangulation, analogously to Direct Linear Transformation (*DLT*) [20]. The cameras Υ are sequentially bundled in the same reference frame. The new ones are stacked according to the L-to-Γ correspondences, computed in the previous stereo pair of cameras. The merged estimations for 3D lines $\{\Gamma_i\}$ are computed as the center of gravity of the spatial lines.

The 3D sketch $\{\Upsilon,\Gamma\}$ generated by linear triangulation is used as input for an optimization algorithm. The least-squares optimization named Sparse Bundle Adjustment (SBA) [14] is based on the Levenberg–Marquardt algorithm, and uses as input the estimated camera extrinsics Υ and the set Γ, now containing unique estimations for each 3D line [21].

The 3D estimations for lines and cameras are drawn in the same spatial sketch, altogether with the cameras. Next, these spatial line segments Γ are fit to different different planes $\mathcal{P}$. Γ is therefore segmented into different groups according to the planes $\mathcal{P}$, and so is done with their projections L. The group of Line Features fitted to the plane $\mathcal{P}_t$ is noted as F_t. The intersections of the coplanar lines F_t on the camera plane Υ^j are the spatial points $\mathcal{T}_t^j$. Therefore, the algorithm can go back to the original images, now known which line segments are coplanar. The intersections of these coplanar lines on the images are described similarly as a feature point. Following this analogy, the descriptor for this feature point will be the pair of two coplanar lines drawing it. We have the correspondences of the straight lines accross images, so we can extrapolate these correspondences to their intersection for the cases where they are coplanar. Secondly, known the correspondences between these intersections, they can be triangulated analogously as it was performed in the first routine with the endpoints of l. The correspondences in $\mathcal{T}_t^j$ are then fed into the linear triangulation algorithm, in order to create initial estimates for the 3D intersections by forward projecting $\mathcal{T}_t^j$. The set of estimations for the 3D points resembling the intersections is a sparse cloud, and it is denoted as $\mathcal{R}$. Finally, and same as with the endpoints, the 3D intersections $\mathcal{R}$ enter the least-squares optimization. The SBA returns the new optimized estimations for Υ, and the optimal 3D intersections $\mathcal{R}$. The spatial line and camera pose estimations are corrected by forward projecting them from the newly estimated camera planes Υ. This returns the final sketch $\{\Upsilon,\Gamma\}$. The high level diagram on **Figure 2** shows the process described in this section.

3.1 Bundle adjustment for line segments

In the case of feature points, the final position of the projected features relative to the camera poses is estimated throughout an optimization process. As a part of most SfM pipelines, bundle adjustment [14] is based on Levenberg-Marquardt, and it rearranges the poses of the cameras and 3D points. The cost function of this optimization process is engineered to find the minimum distance error between the reprojection of every 3D point onto each camera plane and their original observation. A limit value for the residual is usually set to stop the iterative process for the event of convergence, while another threshold is set to end the optimization when reaching a maximum number of iterations.

Along matched segments under a viewpoint change, the only point-to-point valid correspondences are their endpoints. Segment's endpoint location are noticeably less accurate than a rotation and scale invariant feature point. Employing line endpoints as the sole set of geometrical constraints in the adjustment might not be

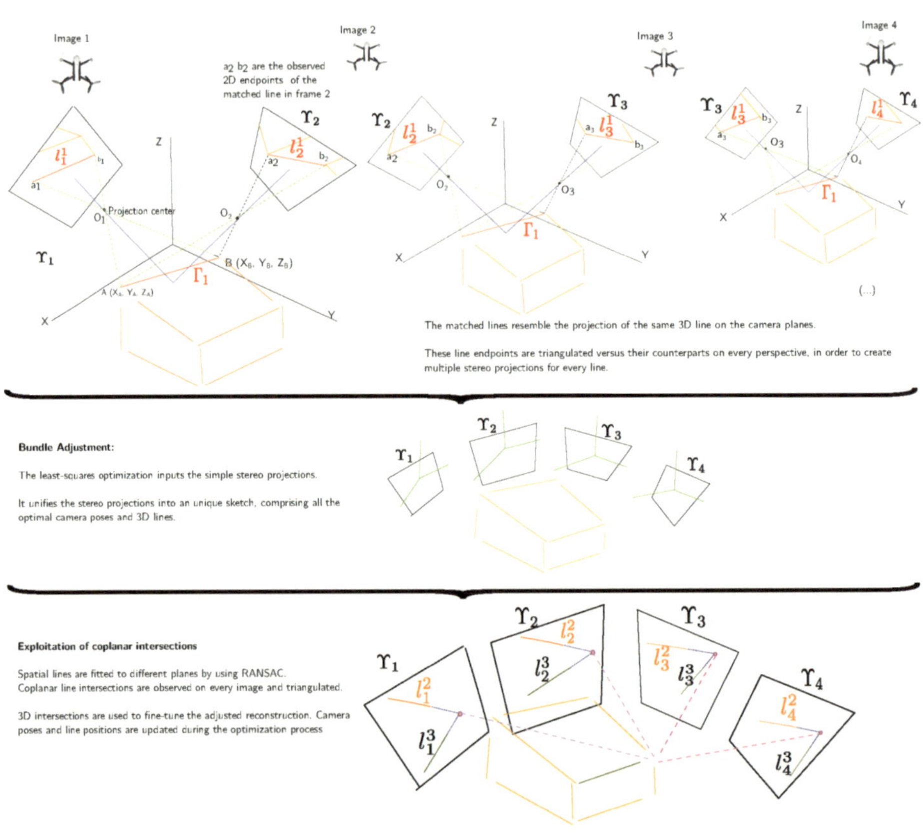

Figure 2.
Figure from [2]. Graphic representation of the 3D abstraction layer of the method. The different cameras are represented as drones.

adequate to improve the 3D sketch. Some of the reasons for this are that recurrent segment mismatches, fragmentation or the inaccurate placement of counterparts may prevent the convergence of the optimization. It is possible to perform a line-based Bundle Adjustment by converting the primitives into Plücker coordinates [20, 21] within the cost function of the optimisation process. This allows a reduction in the number of parameters and the computational cost.

3.2 How to compare the results with Ground Truth meshes

In order to prove the validity of a 3D abstraction method, it has to be benchmarked against a Ground Truth dataset for SfM, which includes both intrinsic and extrinsic parameters for the cameras. These are built with synthetic images from 3D models [22], or with real pictures [23] teamed with 3D model data including the pose of the cameras and the measurements from 3D scanning or Lidar. Both synthetic and real Ground-Truth datasets include a 3D model. The resulting point cloud is aligned with the Ground Truth mesh. The normal distance between the surface of the mesh and the points is computed. In order to assess how the generated sketch fits the Ground Truth model, the Mean Square Error of the distance between both spatial shapes is computed, because it acts as the natural loss function of a Gaussian distribution. In the case of 3D line sketch, in order to compare the sketch with the Ground truth mesh, the 3D straight segments must be discretized into points. To measure the difference in proportions between the generated 3D sketch and the Ground Truth mesh, the normal distance between the surface of the mesh and the discretized points on the lines is computed. Using the obtained error in the distances, discretized points on the lines are coloured to account how far they are from the surface of the mesh. There are several variables that condition the resulting 3D sketch number of images: Firstly, the number of images showing common elements of the scene is one of them. Secondly, the number of segments that can be matched between images. Thirdly, the transformation between both images might condition the matching inlier ratio, and hence, the number of segments correctly projected into space.

For 3D line sketching methods, the length of the final 3D lines will depend on the fragmentation of the detected lines, and its number is closely related to the number of line correspondences between the images. Therefore, results of 3D reconstructions will unavoidably depend on the performance of the method for stages before the spatial projection. Quantitative measurements for 3D abstraction are performed on Ground Truth datasets. The proportions of the generated sketch is measured based on the distance between the segments and the Ground Truth mesh.

Employing a feature-point based abstraction method is profitable for datasets with a sufficient number of pictures featuring textured surfaces, so a dense 3D point cloud can be created. For these 3D abstractions, cameras are located accurately due to the precision of the point rotation and translation invariants. This is the case of the results obtained by abstraction methods working altogether with SIFT pipelines [1, 22], but requiring dozens of high definition pictures with textured surfaces for SIFT to be able to accurately estimate the camera extrinsics.

There are real world applications of Computer Vision that does not always permit to obtain high definition pictures, in textured environments, without blurring and digital noise. For these applications it can be advantageous to estimate the camera extrinsics independently of any feature point 3D reconstruction [2]. **Figure 3** shows a quantitative comparison of the methods [2] and [1] with just 6 and 8 images chosen from the dataset. **Figure 4** increases the number of images to 10 and 12. The test cases are labeled as S_6, comprising image numbers

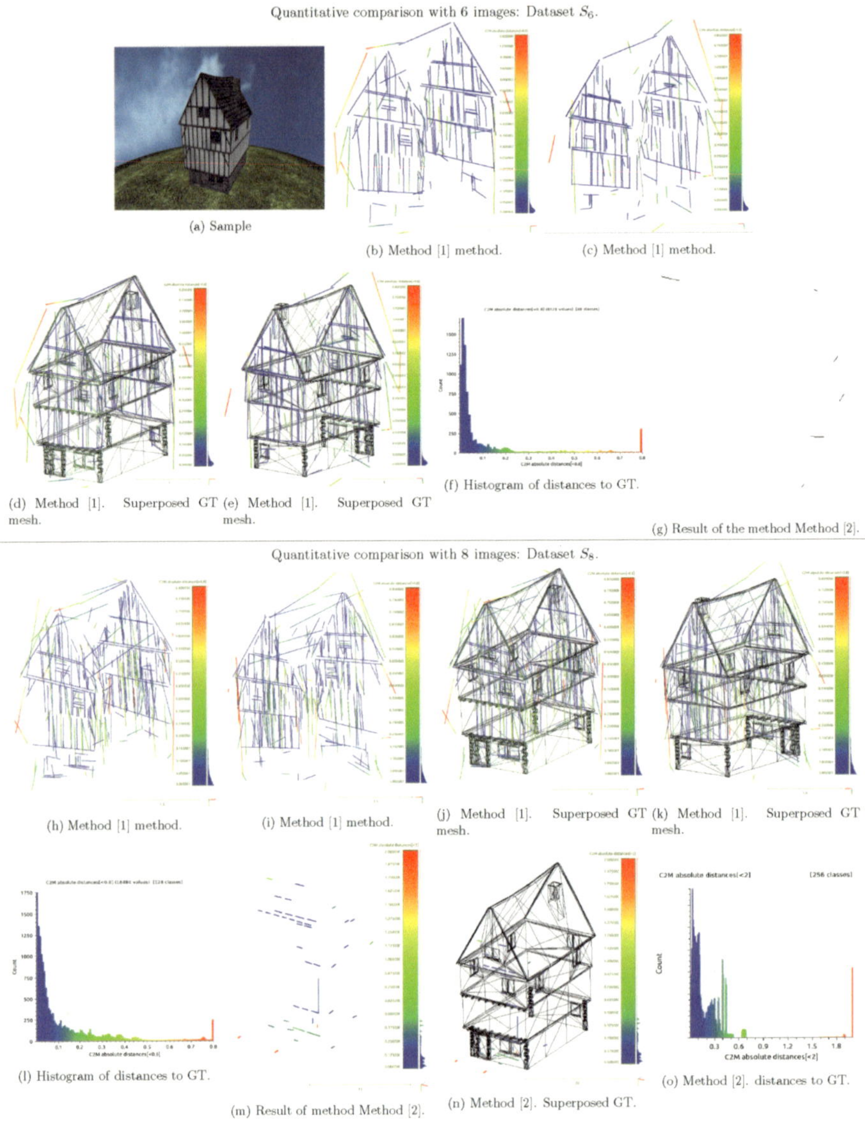

Figure 3.
Figure from [2]. Quantitative comparison using the sets S_6 and S_8. This figure is better viewed on a screen with a 4x zoom. (a) Sample of the set. (b) and (c) [2] against S_6, resulting in 175 lines. (d) and (e) Same superposed onto the Ground Truth mesh. (f) Histogram of distances to Ground Truth with [1] method. The maximum distance to be accounted is set to be 0.8, already considered as outlier. (g) Sparse atomic lines returned by [1] method. (h) to (l) [1] against the set S_8, with 294 segments. (m) and (n) same measurements for the result by [1]. (o) shows the histogram for this latter result.

{6,9,86,46,49,126} from [22], S_8 further add two more images {89,129} to the list, S_{10} includes {8,10,12,88,90,48,50,52,128,130}, and S_{12} further adds images {92,132} to the latter. The resulting 3D line sketches from both sides of the house are aligned by using common lines. This completed sketch is finally aligned to the Ground Truth in order to measure the precision. Note that this experiment takes into account just a the variation of the number of images in the dataset [22]. The results show that the method [2] obtains more usable results for a low number of images, and the results of method [1] are only more adequate than method [2] when the number of images gets close to a dozen. The spatial lines are colored attending to its distance to the surface of the Ground Truth mesh.

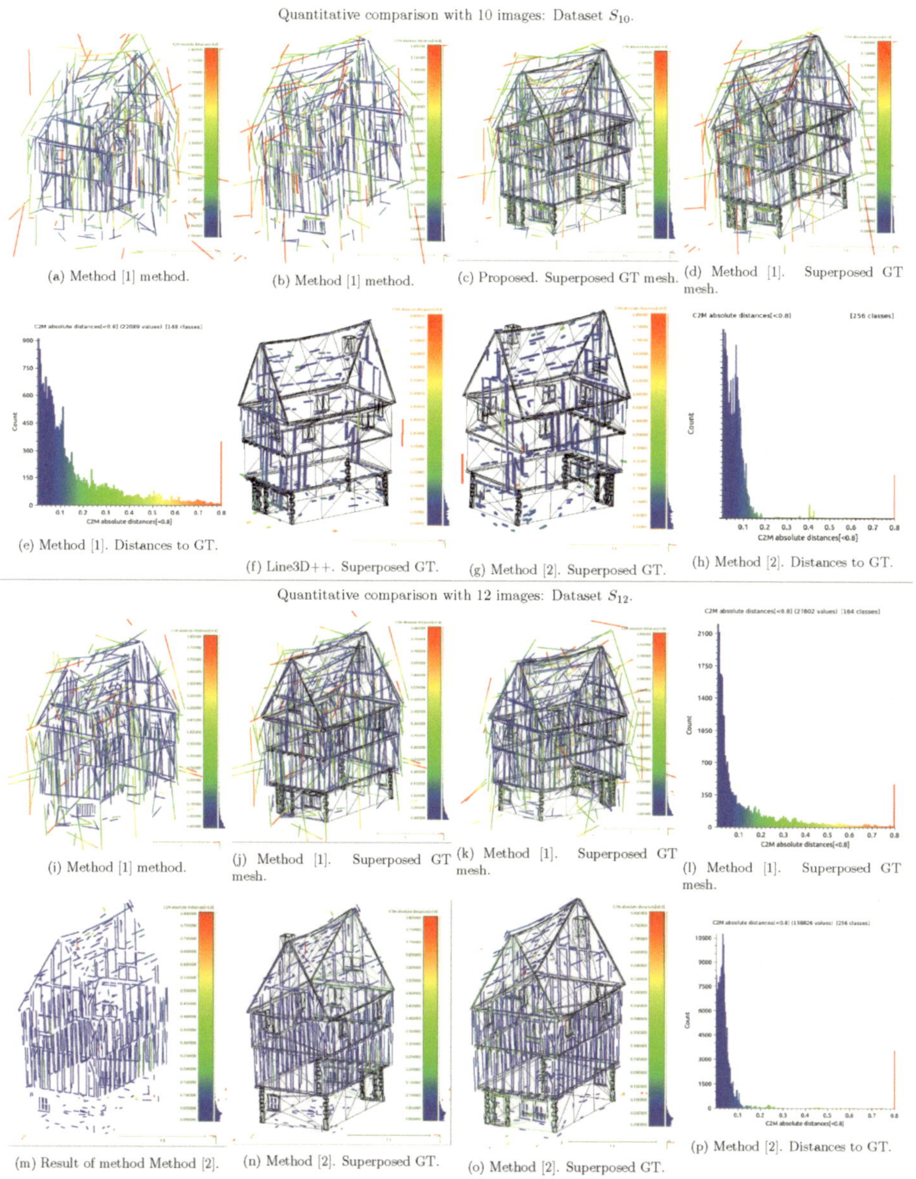

Figure 4.
Figure from [2]. Quantitative comparison using the sets S_{10} and S_{12}. This figure is better viewed on a screen with a 4x zoom. (a), (b) and (c) [2] against S_{10}. The obtained 475 lines have been discretized in points. The distance from each point in the cloud to the surface of the Ground Truth mesh is represented in colors. (d) and (e) Same superposed onto the Ground Truth mesh. (f) Histogram of distances to Ground Truth with the [2] method. The maximum distance to be accounted is set to be 0.8, already considered as outlier. (g) Sparse atomic lines returned by the method [1]. It has been aligned with the Ground Truth mesh. (h) to (l) Same for the method [2] against the set S_{12}, with 556 segments. (m) and (n) same measurements for the result by [1]. (o) shows the histogram for this latter result.

4. Conclusions

A 3D abstraction method receives as input the camera intrinsic parameters and several pictures of the scene. There are two different approaches: The first one does not require the camera extrinsics estimated from an external SfM pipeline, nor the Ground Truth camera poses [2]. It sources the line correspondences from a line matching method, and is able to generate 3D sketches from sets of pictures. This kind of approaches get an edge against datasets with low number of images, or when these present corrupted texture, blurring, and low definition images where the feature point descriptor fails to detect a fair number of keypoints. The reduced number of correspondences limit the thickness of the point cloud generated by the SfM pipelines, and therefore the accuracy of the estimated camera extrinsics. With inaccurate estimations for the cameras, exploiting homography constraints is not adequate to source line correspondences. Oppositely, [2] is able to reconstruct simple line-based sketches with fair precision and number of lines. It required lower number of images to obtain more complete abstractions than method [1]. The range of scenarios where it is advantageous to use method [1] for 3D abstraction includes sets of pictures of simple objects, with low texture, poor illumination, low resolution, blurring or under other conditions that make difficult the success of a point based algorithm. In these scenarios it outperforms the competition in terms of quantity of lines, precision and completeness of the abstraction. Another conclusion is that camera extrinsics are unavoidably required for 3D abstractions featuring many lines, because the estimation for the camera poses will not be accurate if the line matching method returns matching outliers or line fragmentation.

On the other hand, for datasets with moderate number of images, which clear textures, the second approach can be profitable. In this case, the geometric relations from the related points among the images will permit the feature point based pipeline to generate a moderately dense 3D point cloud. In this case, the poses of the cameras obtained by the point based pipeline can be trusted, and used as basis for line matching and linear projection to generate the 3D sketch. The results obtained with method [1] with datasets of hundreds of images are very good. An abstraction using this method will team perfectly with a dense reconstruction.

Both approaches are valid for their range of applications. The first one is valid for difficult datasets with noise and low number of images. The second approach will shine with datasets with high texture and many pictures, because it will will profit of the high precision obtained from feature point based 3D reconstruction pipelines for locating the cameras.

Acknowledgements

The original research work [2] received financial support from the Xunta de Galicia through grant ED431C 2017/69 and Xunta the Galicia (Centro singular de investigación de Galicia accreditation 2016-2019) and the European Union (European Re gional Development Fund - ERDF) through grant ED431G/08.

Conflict of interest

The author declares no conflict of interest.

Author details

Roi Santos Mateos*, Xose M. Pardo and Xose R. Fdez-Vidal
University of Santiago de Compostela, Spain

*Address all correspondence to: roi.santos@usc.es

References

[1] Manuel Hofer, Michael Maurer, and Horst Bischof. Improving sparse 3d models for man-made environments using line-based 3d reconstruction. In 3D Vision (3DV), 2014 2nd International Conference on, volume 1, pages 535–542. IEEE, 2014.

[2] Santos, Roi, Xose M. Pardo, and Xose R. Fdez-Vidal. "Scene wireframes sketching for Unmanned Aerial Vehicles." Pattern Recognition 86 (2019): 354-367.

[3] Zhou, Hang, et al. "Uncertainty Analysis of 3D Line Reconstruction in a New Minimal Spatial Line Representation." Applied Sciences 10.3 (2020): 1096.

[4] Nieto, Marcos, and Luis Salgado. "Real-time robust estimation of vanishing points through nonlinear optimization." Real-Time Image and Video Processing 2010. Vol. 7724. International Society for Optics and Photonics, 2010.

[5] Lowe, David G. "Object recognition from local scale-invariant features." Proceedings of the seventh IEEE international conference on computer vision. Vol. 2. Ieee, 1999.

[6] Alcantarilla, Pablo Fernández, Adrien Bartoli, and Andrew J. Davison. "KAZE features." European Conference on Computer Vision. Springer, Berlin, Heidelberg, 2012.

[7] Micusik, Branislav, and Jana Kosecka. "Piecewise planar city 3D modeling from street view panoramic sequences." 2009 IEEE Conference on Computer Vision and Pattern Recognition. IEEE, 2009.

[8] Criminisi, Antonio, Ian Reid, and Andrew Zisserman. "A plane measuring device." Image and Vision Computing 17.8 (1999): 625-634.

[9] Kim, Hyunwoo, and Sukhan Lee. "Simultaneous line matching and epipolar geometry estimation based on the intersection context of coplanar line pairs." Pattern Recognition Letters 33.10 (2012): 1349-1363.

[10] Langlois, Pierre-Alain, Alexandre Boulch, and Renaud Marlet. "Surface reconstruction from 3d line segments." 2019 International Conference on 3D Vision (3DV). IEEE, 2019.

[11] Santos, Roi, Xose M. Pardo, and Xose R. Fdez-Vidal. "Outlier Detection for Line Matching." Iberoamerican Congress on Pattern Recognition. Springer, Cham, 2018.

[12] Nistér, David. "An efficient solution to the five-point relative pose problem." IEEE transactions on pattern analysis and machine intelligence 26.6 (2004): 756-770.

[13] Fischler, Martin A., and Robert C. Bolles. "Random sample consensus: a paradigm for model fitting with applications to image analysis and automated cartography." Communications of the ACM 24.6 (1981): 381-395.

[14] Triggs, Bill, et al. "Bundle adjustment—a modern synthesis." International workshop on vision algorithms. Springer, Berlin, Heidelberg, 1999.

[15] Zhang, Lilian, and Reinhard Koch. "An efficient and robust line segment matching approach based on LBD descriptor and pairwise geometric consistency." Journal of Visual Communication and Image Representation 24.7 (2013): 794-805.

[16] Wang, Zhiheng, Fuchao Wu, and Zhanyi Hu. "MSLD: A robust descriptor for line matching." Pattern Recognition 42.5 (2009): 941-953.

[17] Bay, Herbert, Vittorio Ferraris, and Luc Van Gool. "Wide-baseline stereo matching with line segments." 2005 IEEE Computer Society Conference on Computer Vision and Pattern Recognition (CVPR'05). Vol. 1. IEEE, 2005.

[18] Lourakis, Manolis IA, Spyros T. Halkidis, and Stelios C. Orphanoudakis. "Matching disparate views of planar surfaces using projective invariants." Image and Vision Computing 18.9 (2000): 673-683.

[19] Jia, Qi, et al. "Novel coplanar line-points invariants for robust line matching across views." European Conference on Computer Vision. Springer, Cham, 2016.

[20] Hartley, Richard, and Andrew Zisserman. Multiple view geometry in computer vision. Cambridge university press, 2003.

[21] Bartoli, Adrien, and Peter Sturm. "Structure-from-motion using lines: Representation, triangulation, and bundle adjustment." Computer vision and image understanding 100.3 (2005): 416-441.

[22] Jain, Arjun, et al. "Exploiting global connectivity constraints for reconstruction of 3D line segments from images." 2010 IEEE Computer Society Conference on Computer Vision and Pattern Recognition. IEEE, 2010.

[23] Strecha, Christoph, et al. "On benchmarking camera calibration and multi-view stereo for high resolution imagery." 2008 IEEE Conference on Computer Vision and Pattern Recognition. Ieee, 2008.

Section 2

Automatic Systems

Chapter 4

Incomplete Data Analysis

Bo-Wei Chen and Jia-Ching Wang

Abstract

This chapter discusses missing-value problems from the perspective of machine learning. Missing values frequently occur during data acquisition. When a dataset contains missing values, nonvectorial data are generated. This subsequently causes a serious problem in pattern recognition models because nonvectorial data need further data wrangling before models are built. In view of such, this chapter reviews the methodologies of related works and examines their empirical effectiveness. At present, a great deal of effort has been devoted in this field, and those works can be roughly divided into two types — Multiple imputation and single imputation, where the latter can be further classified into subcategories. They include deletion, fixed-value replacement, *K*-Nearest Neighbors, regression, tree-based algorithms, and latent component-based approaches. In this chapter, those approaches are introduced and commented. Finally, numerical examples are provided along with recommendations on future development.

Keywords: data imputation, missing value analysis, missing data, data wrangling, data analytics

1. Introduction

With recent development of the Internet of Things (IoT), communication technology, and wireless sensor networks, a huge amount of data is generated every day [1, 2]. It becomes easier to collect data in quantities than before, but new challenges subsequently arise. For example, in manufacturing industries, manufacturers often utilize and deploy thousands of sensors in production lines to monitor the quality of products and to detect possible anomaly or abnormal events. Those sensing data are stored in databases for further analysis. Nonetheless, the collected data are not always perfect. Missing-value entries may appear in databases. When a dataset contains missing values, it is referred to as an incomplete dataset. Missing values in manufacturing industries frequently occur due to sensor failure, particle occlusion, and physical/chemical interferences [3–5]. Unfortunately, most of them happen due to unknown causes and unforeseen circumstances. In addition to manufacturing industries, missing value problems also frequently occur in biomedical areas, for instance, microarray profiling — One of the commonly used tools in genomics [6]. Microarrays are well known for rapidly automated measurement at massive scale — High-throughput biotechnologies [7], but microarrays suffered from missing-value problems [8]. During microarray profiling, missing values might occur as a result of different reasons, such as human errors, dust or scratches on the slide [9], spotting problems, poor/failed hybridization [10], insufficient image resolutions, and fabrication errors [11]. Those unpredictable factors thereby increase the opportunity of defect microarrays. In fact, missing-value problems almost challenge every part of our daily applications ranging from manufacturing/biotechnology industries that rely on sensors to typical service industries that involve questionnaire-based surveys. Questionnaires are used to collect information from respondents. Nonetheless, respondents occasionally fail to provide answers that match the format to fit the response categories [12], subsequently generating unanswered questions or invalid formats/responses (e.g., out-of-range values). There are many reasons for such problems, e.g., respondents refused to answer, respondents chose wrong formats [13, 14], respondents intentionally/unintentionally left blanks, testers addressed unclear/confusing choices, designs involved sensitive/private questions, and interviews were interrupted. These factors could result in missing values in questionnaires.

The difficulty of processing incomplete data is that when a dataset contains missing values, the corresponding entries are marked with invalid values. Accordingly, such a dataset becomes nonvectorial because invalid values are present (which are constantly represented as Not-a-Number (NaN)). To tackle those entries with NaN, mathematical operations (e.g., pairwise distance) need further revision under such circumstances because nonvectorial arithmetic is not well defined.

To handle missing-value problems, data imputation is generally used. Data imputation is a statistical term that describes the process of substituting estimated values for missing ones. Related approaches for data imputation [15] can be classified into two types: Multiple imputation and single imputation. The former is aimed at generating two distributions. One is a distribution for selecting hyperparameters, and the other is a distribution for generating data. Multiple imputation uses a function for generating distributional hyperparameters and takes samples from such a function to obtain an averaged distributional hyperparameter set. Multiple imputation then utilizes this averaged distributional hyperparameter set to create a statistical distribution for describing the data distribution. Finally, data samples are drawn to replace missing values. Popular methods for multiple imputation include Expectation Maximization (EM) algorithms or Monte Carlo Markov Chain (MCMC) strategies [16, 17]. Regarding single imputation, it does not involve drawing data samples from an uncertain function to substitute for missing data as multiple imputation does. In brief, single imputation relies on neither sample drawing nor uncertain functions. At present, a great deal of effort has been devoted to single imputation, for example, hot-deck/cold-deck, deletion, fixed-value replacement (e.g., zeros, means, and medians), K-Nearest Neighbors (KNNs) [18], regression [10, 19, 20], tree-based algorithms [21, 22], and latent component-based approaches (including matrix completion) [15, 23, 24]. This chapter focuses on K-Nearest Neighbors, regression, tree-based algorithms, and latent component-based approaches because the imputation errors of deck/cold-deck and fixed-value replacement are not satisfactory. Moreover, deletion could result in loss of discriminant features or samples. Therefore, the subsequent sections lay emphasis on the other methods.

The rest of this chapter is organized as follows. Section 2 introduces related works, and their methods are subsequently introduced. Sections 3 shows the numerical results, and finally the conclusions are drawn in Section 4.

2. Imputation methods

The following subsections introduce data imputation using KNNs, regression, tree-based algorithms, and latent component-based approaches, respectively. For clarity, the description on the dataset uses the following definitions and notations. A nonmissing-value dataset is represented as $\mathbf{X} = \{\mathbf{x}_n \mid n = 1, 2, \ldots, N\}$, where

$\mathbf{x}_n \in \mathbb{R}^M$, $N \in \mathbb{Z}_+$ refers to the number of samples, and the dimensionality of a sample is $M \in \mathbb{Z}_+$. Herein, the dimensionality represents the number of independent variables, predictor variables, features, dimensions, control variables, or explanatory variables. Those terms are used, depending on various research fields, e.g., data science, machine learning, and statistics. Furthermore, $\mathbf{X}$ is max-min normalized.

If supervised learning is required, the label information and the response variable corresponding to each sample $\mathbf{x}_n$ are respectively defined as c_n and y_n. The former belongs to categorical variables $\in \mathbb{R}$ after encoded, and $\mathbf{c} = \{c_n | n = 1, 2, \dots, N\}$. The latter belongs to numerical variables $\in \mathbb{R}$, and $\mathbf{y} = \{y_n | n = 1, 2, \dots, N\}$. Moreover, the sizes of $\mathbf{X}$, $\mathbf{c}$, and $\mathbf{y}$ are M-by-N, 1-by-N, and 1-by-N. For $\mathbf{X}$, $\mathbf{c}$, and $\mathbf{y}$ that contain missing values, $\tilde{\mathbf{X}}$, $\tilde{\mathbf{c}}$, and $\tilde{\mathbf{y}}$ are used.

2.1 Imputation based on *K*-nearest neighbors

KNNImpute [9] is a popular imputation tool that leverages KNN algorithms to find the K-nearest neighbors nearby a given sample $\tilde{\mathbf{x}}_t$ that contains missing values (if no missing values are present, $\mathbf{x}_t \in \mathbb{R}^M$). The substituted values are generated based on the weighted average of those K-nearest neighbors. Notably, there is a limitation for the selected K-nearest neighbors when KNNImpute is executed. That is, the dimensions of those K-nearest neighbors corresponding to missing-value entries should contain nonmissing-value data. In the plain version, label information was not used while the K-nearest neighbors were searched (**Table 1**).

In the above-mentioned algorithm, ":" means selecting the entire rows or columns based on the position, and the operator $\oplus$ means to replace the missing values with the corresponding generated substituted values. Moreover, distance$(\cdot,\cdot)$ signifies the distance between two samples, e.g., Euclidean or Manhattan distance. Those substituted values were fixed and unchanged once they were generated. Nonetheless, such substituted values were highly affected by initial conditions, such as the subset of M independent variables and the number of nearest neighbors. Iterative K-Nearest Neighbor imputation (IKNNimpute) [25] improved one of such drawbacks by using a loop that iteratively produced substituted values, chose the subset of M independent variables, and reselected near neighbors. **Table 2** lists a simple version of IKNNimpute, where $\tilde{\mathbf{X}}'[j]$ represents the matrix, of which the missing-value entries are filled in with substituted values in the j-th iteration, and J denotes the number of iterations. Besides, $\tilde{\mathbf{X}}'$ is formed by horizontally concatenating $\tilde{\mathbf{X}}$ and $\tilde{\mathbf{X}}_t$. Gray KNNs [26] further proposed Gray Relational Analysis to capture

Algorithm: KNNImpute
Input: $\mathbf{X}$ and $\tilde{\mathbf{x}}_t$
Output: $\hat{\mathbf{x}}_t$
Select M' dimensions ($M' < M$ and $M' \in \mathbb{Z}_+$) without missing values from $\tilde{\mathbf{x}}_t$
Store the indices of the selected dimensions in an M'-by-1 vector $\mathbf{s}$
Apply KNN algorithm to $(\tilde{\mathbf{x}}_t)_{\mathbf{s},:}$ based on the dataset $\mathbf{X}_{\mathbf{s},:}$
Store the K-nearest samples in an M'-by-K matrix $\mathbb{K}$
Compute a K-by-1 weight vector $\mathbf{\Omega} = [1/\text{distance}(k, t) \| k = 1, \dots, K]$
$(\hat{\mathbf{x}}_t)_{\mathbf{s},:} = (\tilde{\mathbf{x}}_t)_{\mathbf{s},:} \oplus (\mathbb{K}\mathbf{\Omega})$

Table 1.
KNNImpute.

Algorithm: IKNNImpute
Input: $\tilde{\mathbf{X}}$ and $\tilde{\mathbf{x}}_t$
Output: $\hat{\mathbf{X}}$ and $\hat{\mathbf{x}}_t$
Form $\tilde{\mathbf{X}}'$ by horizontally concatenating $\tilde{\mathbf{X}}$ and $\tilde{\mathbf{x}}_t$
Form $\tilde{\mathbf{X}}'[1]$ by filling in the missing-value entries of $\tilde{\mathbf{X}}'$ with initial replacement
For $j = 1{:}J$
For $n = 1{:}(N+1)$
Apply KNN algorithm to $\tilde{\mathbf{x}}_n[j]$ based on the dataset $\tilde{\mathbf{X}}'[j]$
Store the K-nearest samples in an M-by-K matrix $\mathbb{K}$
Compare a K-by-1 weight vector $\mathbf{\Omega} = [1/\text{distance}(k,n) \vert k = 1, \ldots, K]$
$\hat{\mathbf{x}}_n[j+1] = \tilde{\mathbf{x}}_n[j] \oplus (\mathbb{K}\mathbf{\Omega})$
End
End

Table 2.
IKNNImpute.

pairwise distance between samples, so that near neighbors were appropriately measured and described.

2.2 Imputation based on regression

The underlying model of this category is primarily based on well-known Ordinary Least Squares (OLS), which focuses on minimizing least squares errors

$$E_{\text{OLS}} = \|\mathbf{y} - \hat{\mathbf{y}}\|_2^2 = \|\mathbf{y} - \mathbf{w}^\top\mathbf{X}\|_2^2 = \text{Tr}\left(\left(\mathbf{y} - \mathbf{w}^\top\mathbf{X}\right)^\top\left(\mathbf{y} - \mathbf{w}^\top\mathbf{X}\right)\right). \quad (1)$$

Herein, $\|\cdot\|_2$ is the $\mathcal{L}_2$-norm distance, $\mathbf{w} \in \mathbb{R}^M$ means an unknown weight vector, T represents matrix transpose, and $\hat{\mathbf{y}} = \mathbf{w}^\top\mathbf{X}$. Moreover, $\mathbf{w} = (\mathbf{X}\mathbf{X}^\top)^{-1}\mathbf{X}\mathbf{y}^\top$ is the closed form for finding the weight vector. Given a nonmissing-value sample $\mathbf{x}_t$, of which the response variable $\tilde{y}_t$ is unknown, $\mathbf{w}^\top\mathbf{x}_t$ can generate estimated results. Such methods included Least Squares Imputation (LSImpute) [10], Local LSImpute (LLSImpute) [19], Sequential LLSimpute (SLLSimpute) [27], Iterated LLSImpute (ILLSImpute) [20], Weighted ILLSImpute [28], Regularized LLSImpute (RLLSImpute) [29], and so on.

In LSImpute [10], the authors examined two types of correlations — those among independent variables (i.e., estimating $\tilde{y}_t$ based on $\mathbf{X}$ and $\mathbf{y}$) and those among samples (i.e., estimating $\tilde{y}_t$ based only on the subsets of $\mathbf{y}$). In the latter, $\tilde{y}_t$ and the selected subsets from $\mathbf{y}$ should be highly correlated during OLS modeling [30]. A combined model was also derived by taking the weighted average from the estimated $\hat{y}_t$ based on independent variables and that based on samples. Unlike LSImpute, LLSImpute [19] was aimed at the correlation among independent variables, but the response variable became multivariate, namely, $\mathbf{Y} = \{Y_n \vert n = 1, 2, \ldots, N\}$, where $Y_n \in \mathbb{R}^L$, and $\mathbf{Y}$ was an L-by-N matrix. The fitting weight was converted from $\mathbf{w}$ to an M-by-L matrix $\mathbf{W}$. For ILLSImpute, different subsets of independent variables in $\mathbf{X}$ were drawn and examined during iterations to estimate an optimal $\hat{y}_t$. SLLSimpute [27] adopted multistage data imputation, where the whole missing-value entries were divided into multiple groups based on the missing rate of each independent variable. The groups with lower missing rates were filled in with substituted values first. Then, those recovered groups were utilized for estimating the other groups with higher missing rates during data reconstruction. For other methods like Weighted ILLSImpute [28] and Regularized LLSImpute (RLLSImpute) [29], variants of OLS models were utilized to adapt to data imputation, for example, Locally Weighted OLS, $\mathcal{L}_2$-norm regularized OLS (i.e., ridge regression), $\mathcal{L}_1$-norm OLS (e.g., LASSO (Least Absolute Shrinkage and Selection Operator), Group LASSO, and Sparse Group LASSO), and regression based on other norms [31].

2.3 Imputation based on tree-based algorithms

Since Random Forests (RFs) [32] were proposed, the performance of tree-structured classification and regression algorithms were significantly enhanced. Random Forests adopted Bagging (i.e., Bootstrap Aggregating) techniques to perform data sampling and perceptron learning. Multiple trees were created based on randomly selected features several times (where the numbers of features should be smaller than M), and each set of sampled data after Bootstrapping was used to train only one tree. A distinct structure used in Random Forests was called proximity matrix, which recorded the concurrence of pairwise samples in a leaf node, namely, the frequencies when two samples coexist in the same leaf node. Such a proximity matrix was symmetric, of which the size was N-by-N, and it was used as imputation weighting during data recovery.

Based on [22], two basic principles for handling incomplete data can be derived by summarizing the strategies mentioned in [22] — Preimputation [33] and on-the-fly imputation [21]. The former filled in missing-value entries with fixed replacement at the beginning. Replacement was iteratively updated using proximity matrices. RF growing process repeated until convergence. For the latter method, on-the-fly imputation did not fill in missing-value entries with initial substituted values, and it skipped using missing-value samples for computing splitting nodes. When missing-value samples reached leaves, imputation started. As assignment of missing-value samples to leaves involved randomness, iterations until convergence improved imputation.

2.4 Imputation based on latent component-based approaches

This type of method has a general procedure for reconstructing an incomplete data matrix. Firstly, the missing-value entries of a data matrix $\tilde{\mathbf{X}}$ are filled in with replacement (e.g., zeros). Secondly, new matrix factors or vector factors are initialized by generating random numbers. In typical, two or three matrix/vector factors are used, e.g., $\mathbf{P} \in \mathbb{R}^{M \times D}$ and $\mathbf{T} \in \mathbb{R}^{D \times N}$. Besides, the product of those factors should be close to $\tilde{\mathbf{X}}$. Thirdly, iterations are performed to improve the replacement. Unlike the aforementioned types of methods, this type has a unique characteristic — Setting the number of latent components $D \in \mathbb{Z}_+$ — before imputation starts.

Popular methods included SVDImpute (i.e., imputation based on Singular Value Decomposition) [34], Nonlinear Iterative Partial Least Squares-Principal Component Analysis (NIPALS-PCA) [35], matrix completion/approximation, and so forth. The common place of SVDImpute and NIPALS-PCA was that projection matrices (or eigenvectors) were computed, whereas plain NMF did not.

2.4.1 SVDImpute

For SVDImpute, let $\hat{\mathbf{X}}[\iota]$ represent the data matrix, of which the missing-value entries are filled in with substituted values in the ι-th iteration. Subsequently, SVD is perform on $\hat{\mathbf{X}}[\iota]$, so that

$$\left(\mathbf{U\Sigma V}^{\top}\right)[\iota] = \mathrm{SVD}\left(\hat{\mathbf{X}}[\iota]\right) \tag{2}$$

where **U** is an M-by-M unitary matrix, $\mathbf{\Sigma}$ represents an M-by-N diagonal matrix, of which the diagonal terms are nonnegative real numbers sorted in descending order, and **V** denotes an N-by-N matrix. By selecting the top D largest diagonal values from $\mathbf{\Sigma}$, the following two corresponding matrices **P** and **T** are formed

$$\begin{cases} \mathbf{P}[\iota] = \mathbf{U}_{:,1:D}[\iota] \\ \mathbf{T}[\iota] = \left(\mathbf{\Sigma V}^{\top}\right)_{1:D,:}[\iota] \end{cases}, \tag{3}$$

where ":,1:D" means selecting columns ranging from the first one to the D-th one, and "1:D,:" extracts rows. The process of (3) is the same as Truncated SVD. Subsequently, the reconstructed data matrix becomes

$$\hat{\mathbf{X}}[\iota+1] = \tilde{\mathbf{X}} \oplus (\mathbf{P}[\iota]\mathbf{T}[\iota]). \tag{4}$$

Herein, the operator $\oplus$ means to replace the missing values of $\tilde{\mathbf{X}}$ with the corresponding generated values by $\mathbf{P}[\iota]\mathbf{T}[\iota]$. Eqs. (2)–(4) iterate until convergence.

2.4.2 NIPALS-PCA

As for NIPALS-PCA (abbreviated as NP below), it minimizes the reconstruction error of

$$\begin{aligned} E_{\mathrm{NP}} &= \sum_{m=1}^{M}\sum_{n=1}^{N}\left(\mathbf{H}\odot\left(\hat{\mathbf{X}}-\mathbf{PT}\right)\right)_{mn}^{2} \\ &= \sum_{n=1}^{N}\left\{\left(\mathbf{H}_{:,n}\odot\left(\hat{\mathbf{X}}_{:,n}-(\mathbf{PT})_{:,n}\right)\right)^{\top}\left(\mathbf{H}_{:,n}\odot\left(\hat{\mathbf{X}}_{:,n}-(\mathbf{PT})_{:,n}\right)\right)\right\} \end{aligned}, \tag{5}$$

where $\odot$ is the elementwise multiplication, **H** denotes an M-by-N index matrix of the nonmissing-value entries in $\tilde{\mathbf{X}}$. That is, if the entry $\tilde{\mathbf{X}}_{m,n}$ is nonmissing, then $\mathbf{H}_{m,n}$ is one. Otherwise, it shows a zero. Let $\mathbf{\Phi} = \mathrm{diag}(\mathbf{H}_{:,n})$ and $\mathbf{\Psi} = \mathrm{diag}(\mathbf{H}_{m,:})$, where n = 1,2, ... ,N, and m = 1,2, ... ,M. Eq. (5) becomes convex if either **P** or **T** is fixed. Then, a solution can be achieved based on Alternating Least Squares (ALS). Taking the derivative form of (5) with respect to $\mathbf{T}_{:,n}$ and zeroing the result yield

$$\mathbf{T}_{:,n} = \left(\mathbf{P}^{\top}\mathbf{\Phi}\mathbf{P}\right)^{-1}\mathbf{P}^{\top}\mathbf{\Phi}\hat{\mathbf{X}}_{:,n}. \tag{6}$$

Likewise, taking the derivative form of (5) with respect to $\mathbf{P}_{m,:}$ and arranging the result generate

$$\mathbf{P}_{m,:} = \hat{\mathbf{X}}_{m,:}\mathbf{\Psi}\mathbf{T}^{\top}\left(\mathbf{T}\mathbf{\Psi}\mathbf{T}^{\top}\right)^{-1}. \tag{7}$$

At the beginning, NIPALS-PCA utilizes the SVD result in the first iteration (see (2)) and extracts the top D components from **U** and **V** as **P** and **T** (see (3)). Subsequently, alternating computation between **P**, **T**, and $\tilde{\mathbf{X}}\oplus(\mathbf{PT})$ until convergence generates solutions.

2.4.3 NMFImpute

Alternating Least Squares has been widely applied to many models, especially matrix completion/approximation. Nonnegative matrix factorization (NMF) is an important topic in matrix completion/approximation, and it became a highlight when it was applied to recommendation systems in a Netflix contest [36–38]. At present, NMF has developed multiple variants, including (i) regularization based on $\mathcal{L}_1$-norms, $\mathcal{L}_2$-norms [39], $\mathcal{L}_{2,1}$-norms, nuclear norms, mixed norms, and graphs, (ii) different loss functions like Huber loss, the correntropy induced metric [40, 41], Cauchy functions [42], and Truncated Cauchy functions [43, 44], and (iii) many more, such as projected gradient NMF [45], projective NMF [46, 47], and orthogonal NMF [48]. The following uses the plain version of NMF as an example to elaborate the detail. NMF minimizes the Least Squares error of

$$E_{\text{NMF}} = \left\|\hat{\mathbf{X}} - \mathbf{PT}\right\|_2^2 = \text{Tr}\left\{\left(\hat{\mathbf{X}} - \mathbf{PT}\right)^{\mathsf{T}}\left(\hat{\mathbf{X}} - \mathbf{PT}\right)\right\}, \tag{8}$$

where $\text{Tr}(\cdot)$ is the trace operator. Eq. (8) is nonconvex and hard to solve. When one variable is fixed, (8) becomes convex. One can use ALS or Coordinate Descend to find the solution. Differentiating (8) with respect to $\mathbf{P}_{m,d}$ and $\mathbf{T}_{d,n}$ (where $d = 1,2, \dots ,D$), respectively, yields

$$\begin{cases} \dfrac{\partial E_{\text{NMF}}}{\partial \mathbf{P}_{m,d}} = 2\left(\mathbf{PTT}^{\mathsf{T}} - \hat{\mathbf{X}}\mathbf{T}^{\mathsf{T}}\right)_{m,d} \\ \dfrac{\partial E_{\text{NMF}}}{\partial \mathbf{T}_{d,n}} = 2\left(\mathbf{P}^{\mathsf{T}}\mathbf{PT} - \mathbf{P}^{\mathsf{T}}\hat{\mathbf{X}}\right)_{d,n} \end{cases}. \tag{9}$$

The multiplicative update rules for (9) are, respectively,

$$\mathbf{P}_{m,d} = \max\left(\varepsilon, \mathbf{P}_{m,d} \odot \left(\frac{\hat{\mathbf{X}}\mathbf{T}^{\mathsf{T}}}{\mathbf{PTT}^{\mathsf{T}}}\right)_{m,d}\right) \tag{10}$$

and

$$\mathbf{T}_{d,n} = \max\left(\varepsilon, \mathbf{T}_{d,n} \odot \left(\frac{\mathbf{P}^{\mathsf{T}}\hat{\mathbf{X}}}{\mathbf{P}^{\mathsf{T}}\mathbf{PT}}\right)_{d,n}\right), \tag{11}$$

where ε is an extremely small positive number, and division is elementwise. Eqs. (10), (11), and (4) iterate until convergence.

3. Experimental results

To show the imputation performance of the above-mentioned methods, experiments on open datasets were conducted. The datasets included Abalone (Aba), Scene (SCN), White Wine (WW), and Indian Pines (IP). The number of samples were 4177, 2407, 4898, and 21025, respectively. Furthermore, the dimensionality was, respectively, 561, 294, 11, and 200. Imputation approaches included KNN Regression Imputation (KNRImpute), KNNImpute with $K = 5$, Regression Tree Imputation (RTImpute), Random Forest Imputation (RFImpute), and NIPALS-PCA Imputation (PCAImpute) with only one component. All of them were found in open sources.

To generate missing values for each dataset, this study used a random generator to decide missing-values entries. For KNRImpute, KNNImpute, RTImpute, and RFImpute, they required that missing values should not be uniformly distributed in data. Otherwise, imputation could not be performed. Thus, not every of the independent variables were chosen. Missing-value rates ranged from 3.00% to 9.00%, with a separation of 2.00%. When a dataset was recovered, the difference between the substituted values and the ground truth was compared. The criteria for examining the quality of imputation included root-mean-squared errors (RMSEs) and coefficients of determination (R^2). For coefficients of determination, this study reshaped (i.e., vectorized) a dataset into a vector and then used the following definition to compute R^2. Assume that x_g represents an element of a ground-truth dataset ($g = 1, \dots, MN$), $\hat{x}_g$ denotes the corresponding recovered value, and $\overline{x}_g$ denotes the mean of all the ground-truth values in the same dataset

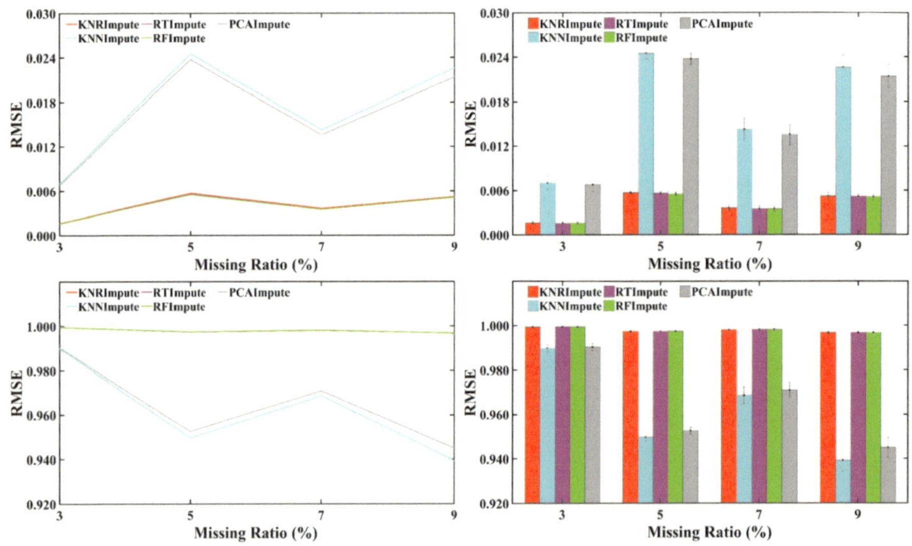

Figure 1.
Average and the standard deviation of the RMSEs and R^2 based on dataset Aba. The top ones are RMSEs, and the bottom are R^2. The standard deviation was divided by 10.000 for better resolutions.

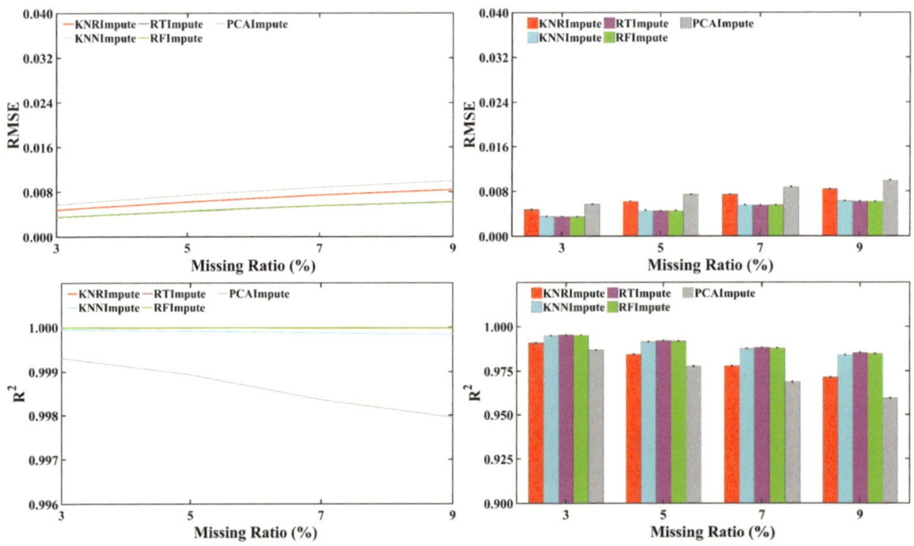

Figure 2.
Average and the standard deviation of the RMSEs and R^2 based on dataset SCN. The top ones are RMSEs, and the bottom are R^2.

$$R = 1 - \sum_g (x_g - \hat{x}_g)^2 / \sum_g (x_g - \overline{x}_g)^2. \tag{12}$$

When R was close to one, substituted values approached the ground truth. This implied that the difference between the substituted values and the ground truth was smaller.

Figures 1–4 display the average and the standard deviation of the RMSEs and R^2, where the horizontal axis denotes the missing rates, and the vertical axis is the evaluation result. The left subplots are line plots, and the right ones show bar charts with standard variation. As shown in the figures, standard variation was quite small.

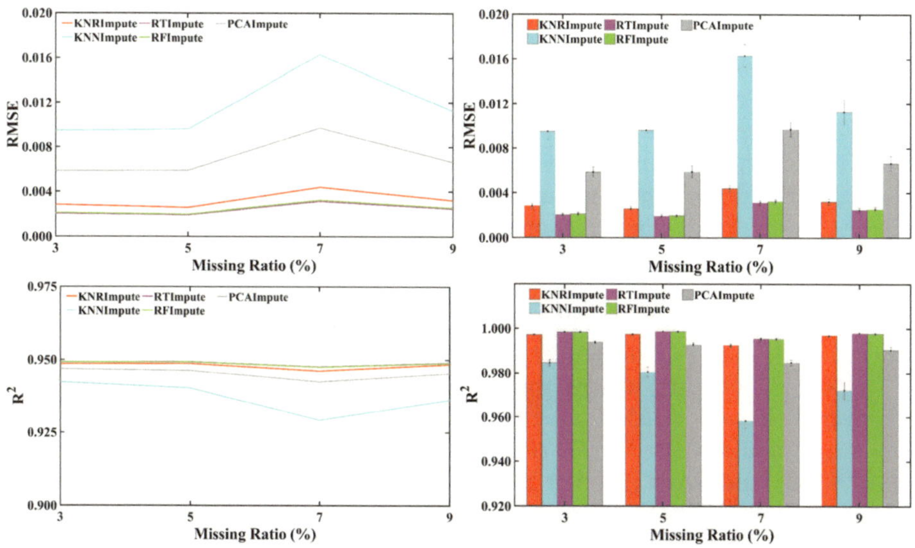

Figure 3.
Average and the standard deviation of the RMSEs and R^2 based on dataset WW. The top ones are RMSEs, and the bottom are R^2. The standard deviation was divided by 10.000 for better resolutions.

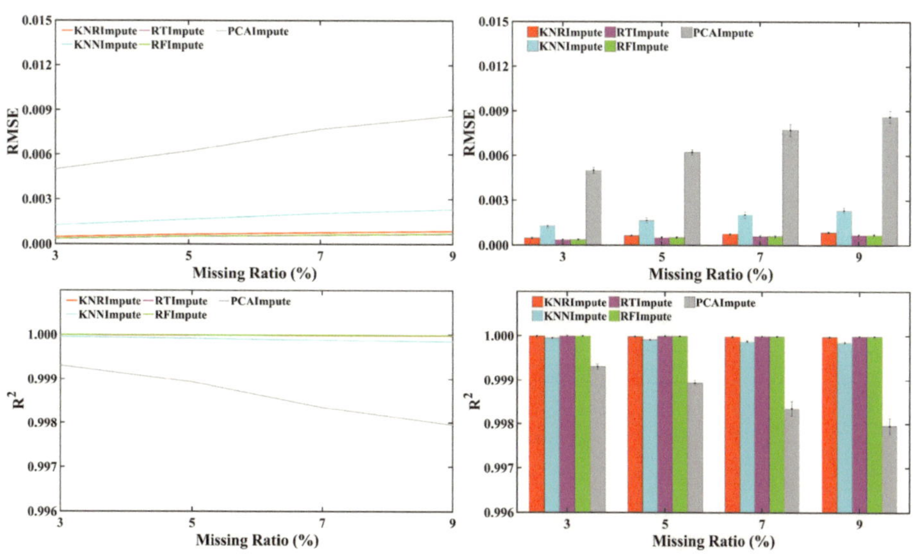

Figure 4.
Average and the standard deviation of the RMSEs and R^2 based on dataset IP. The top ones are RMSEs, and the bottom are R^2.

Besides, RMSEs became higher when missing rates were increased. Observations showed that KNRImpute, RTImpute, and RFImpute generated similar RMSEs. Overall, KNNImpute and PCAImpute were affected by the hyperparameters.

4. Conclusions

This chapter introduces recent methods for processing missing values. Besides, four types of commonly used algorithms, namely, *K*-Nearest Neighbors, regression, tree-based algorithms, and latent component-based approaches, were examined. Their advantages and disadvantages were also discussed in each subsection. It is worth noting that data imputation usually does not require training data. It becomes impractical when data imputation needs supervisory information or the ground truth (notably, the ground truth is unobservable). This is because when missing values occur in training data and even when the ground truth is missing, the supervised methods even cannot work to learn the ground truth. Therefore, those selected four types of commonly used algorithms in this chapter did not rely on and require any supervisory information.

To evaluate those commonly used algorithms, this chapter conducted experiments on open datasets. Criteria including root-mean-squared errors and coefficients of determination were adopted. Numerical results were also displayed in the experimental section for reference.

In more recent years, surveys showed that a deep learning model "Generative Adversarial Network (GAN)" has attracted much attention, and several novel imputation methods based on GANs have been proposed, e.g., MisGAN [49], MIWAE [50], and GAIN [51]. For future studies, deep learning architectures such as Deep PCA, PCANet, and Deep NMF, can be integrated into those four types of commonly used algorithms, namely, *K*-Nearest Neighbors, regression, tree-based algorithms, and latent component-based approaches and subsequently enhance data imputation.

Acknowledgements

The authors would like to appreciate Guan-Yu Huang for conducting experiments to collect results. This work is supported in part by Pervasive Artificial Intelligence Research (PAIR) Labs, Taiwan.

DOI: http://dx.doi.org/10.5772/intechopen.94068

Author details

Bo-Wei Chen[1] and Jia-Ching Wang[2*]

1 Department of Electrical Engineering, National Sun Yat-sen University, Kaohsiung City, Taiwan

2 Department of Computer Science and Information Engineering, National Central University, Taoyuan City, Taiwan

*Address all correspondence to: jcw@csie.ncu.edu.tw

References

[1] J. Reilly, S. Dashti, M. Ervasti, J. D. Bray, S. D. Glaser, and A. M. Bayen, "Mobile phones as seismologic sensors: Automating data extraction for the iShake system," *IEEE Transactions on Automation Science and Engineering*, vol. 10, no. 2, pp. 242–251, Apr. 2013.

[2] Z. Peng, S. Gao, B. Xiao, S. Guo, and Y. Yang, "CrowdGIS: Updating digital maps via mobile crowdsensing," *IEEE Transactions on Automation Science and Engineering*, vol. 15, no. 1, pp. 369–380, Jan. 2018.

[3] L. Xu and Q. Huang, "EM estimation of nanostructure interactions with incomplete feature measurement and its tailored space filling designs," *IEEE Transactions on Automation Science and Engineering*, vol. 10, no. 3, pp. 579–587, Jul. 2013.

[4] D. Li, Y. Zhou, G. Hu, and C. J. Spanos, "Handling incomplete sensor measurements in fault detection and diagnosis for building HVAC systems," *IEEE Transactions on Automation Science and Engineering*, vol. 17, no. 2, pp. 833–846, Apr. 2020.

[5] S.-K. S. Fan, C.-Y. Hsu, D.-M. Tsai, F. He, and C.-C. Cheng, "Data-driven approach for fault detection and diagnostic in semiconductor manufacturing," *IEEE Transactions on Automation Science and Engineering*, vol. 17, no. 4, pp. 1925–1936, Oct. 2020

[6] M. M. Babu, "An introduction to microarray data analysis," in *Computational Genomics: Theory and Application*, R. P. Grant, Ed. UK: Taylor & Francis, 2004, pp. 225–249.

[7] T. Aittokallio, "Dealing with missing values in large-scale studies: Microarray data imputation and beyond," *Briefings in Bioinformatics*, vol. 11, no. 2, pp. 253–264, Mar. 2009.

[8] M. S. B. Sehgal, I. Gondal, and L. S. Dooley, "Collateral missing value imputation: A new robust missing value estimation algorithm for microarray data," *Bioinformatics*, vol. 21, no. 10, pp. 2417–2423, Feb. 2005.

[9] O. Troyanskaya, M. Cantor, G. Sherlock, P. Brown, T. Hastie, R. Tibshirani, D. Botstein, and R. B. Altman, "Missing value estimation methods for DNA microarrays," *Bioinformatics*, vol. 17, no. 6, pp. 520–525, 2001.

[10] T. H. Bø, B. Dysvik, and I. Jonassen, "LSimpute: Accurate estimation of missing values in microarray data with least squares methods," *Nucleic Acids Research*, vol. 32, no. 3, Feb. 2004.

[11] C.-C. Chiu, S.-Y. Chan, C.-C. Wang, and W.-S. Wu, "Missing value imputation for microarray data: A comprehensive comparison study and a web tool," *BMC Systems Biology*, vol. 7, no. 6, Dec. 2013.

[12] E. D. de Leeuw, "Reducing missing data in surveys: An overview of methods," *Quality & Quantity*, vol. 35, no. 2, pp. 147–160, May 2001.

[13] S. Dorofeev, *Statistics for Real-life Sample Surveys: Non-simple-random Samples and Weighted Data*. Cambridge, UK: Cambridge University Press, Jul. 2006.

[14] K. P. Soman, S. Diwakar, and V. Ajay, *Insight into Data Mining: Theory and Practice*. India: Prentice-Hall, Apr. 2010.

[15] D. Bertsimas, C. Pawlowski, and Y. D. Zhuo, "From predictive methods to missing data imputation: An optimization approach," *Journal of Machine Learning Research*, vol. 18, no. 1, pp. 7133–7171, Jan. 2017.

[16] Y. C. Yuan, "Multiple imputation for missing data: Concepts and new development," SAS Institute Incorporation, Rockville, MD, Technical Report, 2000.

[17] T. H. Lin, "A comparison of multiple imputation with EM algorithm and MCMC method for quality of life missing data," *Quality & Quantity*, vol. 44, no. 2, pp. 277–287, Feb. 2010.

[18] D. I. Gad and B. R. Manjunatha, "Performance evaluation of predictive models for missing data imputation in weather data," in *Proc. 2017 International Conference on Advances in Computing, Communications and Informatics*, Udupi, India, 2017, Sep. 13–16, pp. 1327–1334.

[19] H. Kim, G. H. Golub, and H. Park, "Missing value estimation for DNA microarray gene expression data: Local least squares imputation," *Bioinformatics*, vol. 21, no. 2, pp. 187–198, Aug. 2004.

[20] Z. Cai, M. Heydari, and G. Lin, "Iterated local least squares microarray missing value imputation," *Journal of Bioinformatics and Computational Biology*, vol. 4, no. 5, pp. 935–957, Oct. 2006.

[21] H. Ishwaran, U. B. Kogalur, E. H. Blackstone, and M. S. Lauer, "Random survival forests," *Annals of Applied Statistics*, vol. 2, no. 3, pp. 841–860, Sep. 2008.

[22] F. Tang and H. Ishwaran, "Random forest missing data algorithms," *Statistical Analysis and Data Mining*, vol. 10, no. 6, pp. 363–377, 2017.

[23] T. Marwala, *Computational Intelligence for Missing Data Imputation, Estimation, and Management: Knowledge Optimization Techniques*. Pennsylvania, USA: Information Science Reference, Mar. 2009.

[24] C. M. Salgado, C. Azevedo, H. Proença, and S. M. Vieira, "Missing data," in *Secondary Analysis of Electronic Health Records*, M. C. Data, Ed. Cham, Switzerland: Springer, 2016.

[25] L. P. Brás and J. C. Menezes, "Improving cluster-based missing value estimation of DNA microarray data," *Biomolecular Engineering*, vol. 24, no. 2, pp. 273–282, Jun. 2007.

[26] S. Zhang, "Nearest neighbor selection for iteratively kNN imputation," *Journal of Systems and Software*, vol. 85, no. 11, pp. 2541–2552, Nov. 2012.

[27] X. Zhang, X. Song, H. Wang, and H. Zhang, "Sequential local least squares imputation estimating missing value of microarray data," *Computers in Biology and Medicine*, vol. 38, no. 10, pp. 1112–1120, Oct. 2008.

[28] Z. Yu, T. Li, S.-J. Horng, Y. Pan, H. Wang, and Y. Jing, "An iterative locally auto-weighted least squares method for microarray missing value estimation," *IEEE Transactions on NanoBioscience*, vol. 16, no. 1, pp. 21–33, Jan. 2017.

[29] A. Wang, Y. Chen, N. An, J. Yang, L. Li, and L. Jiang, "Microarray missing value imputation: A regularized local learning method," *IEEE/ACM Transactions on Computational Biology and Bioinformatics*, vol. 16, no. 3, pp. 980–993, May–Jun. 2019.

[30] M. S. Sehgal, I. Gondal, and L. Dooley, "Gene expression imputation techniques for robust post genomic knowledge discovery," in *Computational Intelligence in Medical Informatics*, A. Kelemen, A. Abraham, and Y. Liang, Eds. Berlin, Heidelberg, German: Springer-Verlag, 2009, pp. 185–206.

[31] X. Chen, Y. Cai, Q. Liu, and L. Chen, "Nonconvex l_p-norm regularized sparse self-representation for traffic sensor

data recovery," *IEEE Access*, vol. 6, pp. 24279–24290, 2018.

[32] L. Breiman, "Random forests," *Machine Learning*, vol. 45, no. 1, pp. 5–32, Oct. 2001.

[33] D. J. Stekhoven and P. Bühlmann, "MissForest—Non-parametric missing value imputation for mixed-type data," *Bioinformatics*, vol. 28, no. 1, pp. 112–118, 2011.

[34] K. A. Severson, M. C. Molaro, and R. D. Braatz, "Principal component analysis of process datasets with missing values," *Processes*, vol. 5, no. 3, pp. 38–55, Jul. 2017.

[35] B. Grung and R. Manne, "Missing values in principal component analysis," *Chemometrics and Intelligent Laboratory Systems*, vol. 42, nos. 1–2, pp. 125–139, Aug. 1998.

[36] A. Paterek, "Improving regularized singular value decomposition for collaborative filtering," in *Proc. 13th ACM SIGKDD International Conference on Knowledge Discovery and Data Mining*, San Jose, California, United States, 2007, Aug. 12–15, pp. 39–42.

[37] X. Kong, F. Xia, J. Wang, A. Rahim, and S. K. Das, "Time-location-relationship combined service recommendation based on taxi trajectory data," *IEEE Transactions on Industrial Informatics*, vol. 13, no. 3, pp. 1202–1212, Jun. 2017.

[38] C. Hsu, M. Yeh, and S. Lin, "A general framework for implicit and explicit social recommendation," *IEEE Transactions on Knowledge and Data Engineering*, vol. 30, no. 12, pp. 2228–2241, Dec. 2018.

[39] Y. Zhou, D. Wilkinson, R. Schreiber, and R. Pan, "Large-scale parallel collaborative filtering for the Netflix prize," in *Proc. 4th International Conference on Algorithmic Applications in Management*, Shanghai, China, 2008, Jun. 23–25, pp. 337–348.

[40] L. Du, X. Li, and Y.-D. Shen, "Robust nonnegative matrix factorization via half-quadratic minimization," in *Proc. 12th IEEE International Conference on Data Mining*, Brussels, Belgium, 2012, Dec. 10–13, pp. 201–210.

[41] R. He, L. Wang, Z. Sun, Y. Zhang, and B. Li, "Information theoretic subspace clustering," *IEEE Transactions on Neural Networks and Learning Systems*, vol. 27, no. 12, pp. 2643–2655, Dec. 2015.

[42] A. Liutkus, D. Fitzgerald, and R. Badeau, "Cauchy nonnegative matrix factorization," in *Proc. 2015 IEEE Workshop on Applications of Signal Processing to Audio and Acoustics*, New Paltz, New York, United States, 2015, Oct. 18–21.

[43] N. Guan, T. Liu, Y. Zhang, D. Tao, and L. S. Davis, "Truncated Cauchy non-negative matrix factorization," *IEEE Transactions on Pattern Analysis and Machine Intelligence*, vol. 41, no. 1, pp. 246–259, Jan. 2019.

[44] H. Wang, W. Yang, and N. Guan, "Cauchy sparse NMF with manifold regularization: A robust method for hyperspectral unmixing," *Knowledge-Based Systems*, vol. 184, Nov. 2019.

[45] C.-J. Lin, "Projected gradient methods for nonnegative matrix factorization," *Neural Computation*, vol. 19, no. 10, pp. 2756–2779, Oct. 2007.

[46] Z. Yuan and E. Oja, "Projective nonnegative matrix factorization for image compression and feature extraction," in *Proc. Scandinavian Conference on Image Analysis*, Joensuu, Finland, 2005, Jun. 19–22, pp. 333–342.

[47] Z. Yang and E. Oja, "Linear and nonlinear projective nonnegative matrix

factorization," *IEEE Transactions on Neural Networks*, vol. 21, no. 5, pp. 734–749, May 2010.

[48] J. Wen, J. E. Fowler, M. He, Y.-Q. Zhao, C. Deng, and V. Menon, "Orthogonal nonnegative matrix factorization combining multiple features for spectral–spatial dimensionality reduction of hyperspectral imagery," *IEEE Transactions on Geoscience and Remote Sensing*, vol. 54, no. 7, pp. 4272–4286, Jul. 2016.

[49] S. C.-X. Li, B. Jiang, and B. M. Marlin, "MisGAN: Learning from incomplete data with generative adversarial networks," in *Proc. 7th International Conference on Learning Representations*, New Orleans, Louisiana, USA, 2019, May 06–09.

[50] P.-A. Mattei and J. Frellsen, "MIWAE: Deep generative modelling and imputation of incomplete data sets," in *Proc. 36th International Conference on Machine Learning*, Long Beach, California, USA, 2019, Jun. 09–15, pp. 4413–4423.

[51] J. Yoon, J. Jordon, and M. v. d. Schaar, "GAIN: Missing data imputation using generative adversarial nets," in *Proc. 35th International Conference on Machine Learning*, Stockholm, Sweden, 2018, Jul. 10–15, pp. 5675–5684.

Chapter 5

Retina Recognition Using Crossings and Bifurcations

Lukáš Semerád and Martin Drahanský

Abstract

Recognition of people on the basis of biometric characteristics has been known for many centuries. One of the most used biometric features is fingerprint. Recently, we have also come across the iris pattern more often. Retinal recognition offers similarly reliable mechanisms, but they are not yet well explored. Our procedure for obtaining a biometric pattern is partly based on fingerprints. In comparison with fingerprints, retinal recognition identifies *bifurcations* or optical *crossings*, i.e., instead of papillary lines, the vessels are used. The procedure is more complicated due to the multiple layers in which the blood vessels intersect. Our work deals with determining the probabilities for various areas of the retina in which bifurcation and crossing occur. It also describes how recognition can be affected by various diseases.

Keywords: Biometrics, human eye, human retina, biometric recognition, blood-vessel bifurcation, blood-vessel crossing, retina imaging, image processing, biometric entropy

1. Introduction

Biometric systems are an increasingly used today, whether it is access control systems or person recognition. Following the widely used biometrics of fingerprints, the use of the iris of the eye is increasing. The retina of the human eye is another possibility in biometrics, but it has not yet been explored as much as the previously mentioned features.

For various applications, it is necessary to create a reliable recognition system based on the given parameters. The main goal of this work is not to create a perfect application for determining the degree of agreement of two retinas, but rather to outline the principles for the individual components of recognition.

The work in the first chapter summarizes the theoretical foundations of the eye and retina. The formation of blood vessels in the vascular bed of the retina is also described here, and the entropic similarity of all retinas is indicated. Finally, tools for examining the eye and for recording a retinal image are also mentioned. The second chapter describes the biometric basis for retinal feature recognition. Existing principles are mentioned here. The following chapter presents the details of the individual parts of our established recognition principle. The used databases and the created programs are presented. The last chapter summarizes the results of all created programs.

IntechOpen

2. The retina

2.1 Eye anatomy

Human eye consists of many working parts such as sclera, cornea, pupil, lens, iris, ciliary body, retina, optic nerve, choroid, etc. *Sclera*, the white colored outer layer of the eye, works as a protector of the eye. *Cornea*, a transparent circular part, refracts the light entering the eye onto the lens. *Lens* is a crystalline part located directly behind the pupil. Its task is to focus light onto the retina. *Pupil* is the dark spot at the center of a healthy iris. It acts as the shutter of a camera since the amount of light entering the human eye is regulated by the diameter of the pupil. *Iris* is the colored, visible part of the eye located in front of the lens. It regulates the amount of light entering the eye by widening (dilation) and narrowing (constriction) of the pupil. *Ciliary body* delivers oxygen and nutrients to the lens and cornea. It contains the ciliary muscle, which changes the shape of the lens when our eyes focus on an object. *Optic nerve* transfers all the visual information from the retina to the brain. *Choroid* is a thin vascular layer between the retina and the sclera. It provides oxygen and nourishment to the outer layers of the retina. It also contains a pigment that absorbs excess light (**Figure 1**) [2].

2.2 Anatomy of the retina

The retina is located at the back of the eye and is the only part of the central nervous system that is non-invasively observable. It is responsible for light rays sensing and vision at all. It is a very thin light-sensitive layer that is transparent. Retinal cameras capture deeper layers of the eye behind the retinal layer. Its thickness is 0.2-0.4 mm and contains two types of light-sensitive cells, rods and cones. The rods, which are approximately 75 to 150 million [3, 4], are used for more sensitive vision in low light. However, the rods only send information to the brain in the grayscale. On the contrary, the cones perceive incident light rays in color, but they need more light to function than rods. There are approximately 7 million of them in one eye and are divided according to the perception of colors red, green, and blue. The peripheral part of the retina is rod-dominated, whereas the yellow spot is cone-dominated. Low density of rods in the yellow spot makes it less sensitive to the light. Retina would be compared with 157-megapixel camera [3, 4].

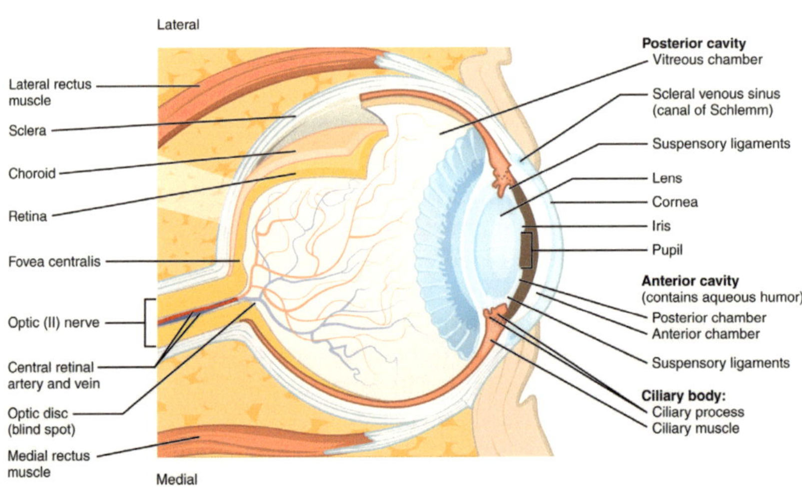

Figure 1.
Anatomy of an eye [1].

Two main parts of the retina are shown in **Figure 2**. It is a blind spot (optic disk) and a macula (yellow spot). Optic disk is a circular area with an average surface of about 3 mm^2 where ganglion cells, i.e., projection neurons transferring information from the retinal neurons, form the optic nerve, the central retinal artery comes into the retina and the retinal vein leaves the retina. The color of a normal optic disk varies from orange to pink. The optic disk is known as a "blind spot" since it does not contain photoreceptors, i.e., rods and cones. Corresponding parts of the visual field are not visible to a person. However, thanks to the ability of the brain to ignore or interpolate missing information from the other eye, we usually do not notice the blind spot. Opposite to the optic disk macula (yellow spot) is the sharpest vision area with a diameter of around 5 mm. The highest concentration of cones on the macula makes it responsible for the perception of colors. Fovea is located in the center of the macula with the densest concentration of photoreceptors in the eye. The image of the object falling on the macula (yellow spot) is reflected in the fovea.

2.3 Retinal vasculature

The structure of retinal vessels is much like the brain and remains unchanged throughout life. Two main sources of blood supplying the retina are the retinal artery and the vessels. The blood vessels nourish the retina's outer layer with photoreceptors. Meanwhile, four major branches of the retinal artery called terminal arterioles, provide blood supply to the retina, nourishing in the first place the inside of it.

One type of retinal scanners is the fundus camera, which is a specialized low power microscope with an attached camera. **Figure 3** shows the fundus camera used in the biometric laboratory of Faculty of Information Technology (FIT) at Brno University of Technology (BUT) in Czech Republic (CZ). No matter which kind of retinal scanners is used, without user's conscious cooperation the retinal image acquisition is not possible.

Blood vessels coming out from the blind spot and make a tree shape on the surface of the retina (shown in **Figure 2**). This tree shape hardly ever changes over the lifetime of an individual except due to some severe eye diseases such as hard glaucoma and cataracts. It is not affected by external environment, since the retina is not an external organ such as a fingerprint [8, 9]. Moreover, it differs from person to person due to many factors such as the thickness of vessels, the distance from

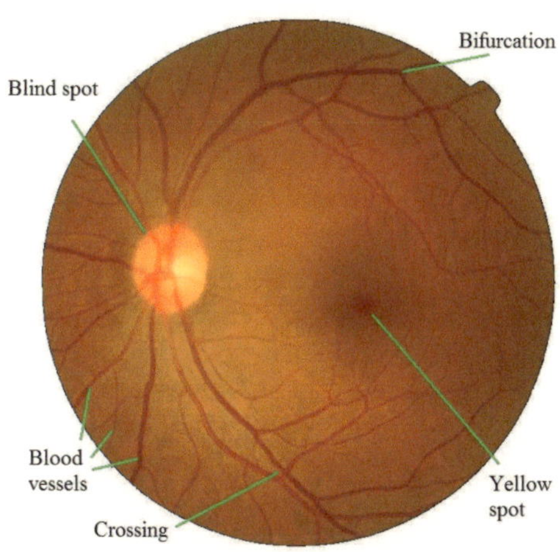

Figure 2.
Retinal image [5].

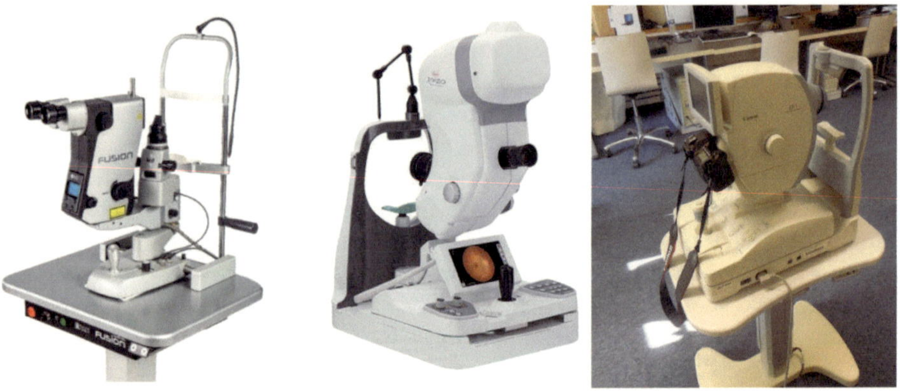

Figure 3.
Slit lamp example (left) [6]; example of a non-mydriatic fundus camera (middle) [7]; (right) fundus camera used in the biometric laboratory at FIT BUT (right).

each other, the presence of bifurcations (division points of a single vessel), crossings (intersection points of two or multiple vessels), ending points of vessels etc., which are all in different locations and in various numbers.

2.3.1 Medicine basis of vein structure

Retinal vasculature occurs mainly by *angiogenesis*. Its formation and the factors that regulate the development of superficial retinal vascular plexus in humans are already quite known. The cell–cell signaling that occurs between different cellular components affects the regression of the vessels, the sprouting angiogenesis, the vascular remodeling, and vessel differentiation events that are involved. These cellular components include neurons, glia, endothelial cells, pericytes, and immune cells.

An invasion of migrating astrocytes coming from the optic nerve and going into the retina precedes the development of the retinal vasculature. They begin from the optic nerve head and extend to the retina's inner surface in a centrifugal fashion as a cell population that is proliferating. They form a cellular mesh-like network which provides the blood vessels a template in their wake. Astrocytes experience hypoxia and express the vascular endothelial growth factor (VEGF) strongly before blood vessels cover it. VEFG is the key stimulus for angiogenesis. It induces the endothelial cells' migration and the nascent vascular network expansion over then retina's inner surface. Bit by bit, the astrocytes start to downregulate the VEGF expression right after a perfused vascular network has formed. There emerges a typical stellate morphology of the retinal blood vessels. Sprouting angiogenesis are also believed to have formed the deeper networks of the retinal vasculature where new vessel sprouts are formed by proliferating endothelial cells and the vascular network is extended from pre-existing vessels [1].

The presence of astrocytes is related to the blood vessels in the retina. Retinal astrocytes and blood vessels are covering the entire retina in primates, except for the primate fovea, which does not have retinal astrocytes and blood vessels. Combined together, these observations imply that the retinal astrocyte network and the retinal vasculature are linked developmentally and are evolutionary.

The process of *retinal angiogenesis* explains why the pattern of retinal vascular network appears quite uniform in the population. Near the disc, the arterioles are more heavily concentrated in the outer choroid, especially nasal and temporal to the disc. The choroidal arteries are wavy or rippled, some like corkscrews, some with one or more 360° loops, and many with rather tightly twisted S-shaped turns.

When the outer vessels are gently removed, the smaller vessels are visible. Posteriorly, most prominently in the submuscular region, the vessels in the middle layers are highly complicated. Branches are usually not equal in length. The angles formed by two twigs range from 30° or 40° to 180°, and T-shaped branchings are common. After the bifurcation, a branch may continue in a straight path, making a sweeping C-shaped curve of 240° to almost 360° and diving inward to enter capillaries only a short distance from the parent trunk.

Most of the bifurcations of the larger choroidal arteries are dichotomous and the very acute angle formed by the offshoots points toward the disc. Second and third branchings may occur almost immediately so that a parent vessel seems to break up, fanlike, into four or six radiating branches after the first branching. Anastomoses between the larger choroidal arteries are not common but are frequent in the smaller branches [10].

2.4 Tools for medical examination and biometric scanning

An image is acquired in the retina, like how a camera captures it. The beam first passes through the pupil of the eye then appears in the focus of the lens on the retina, resembling a film. Specialized optical devices are used in medicine for the visual examination of the retina.

We will first describe the existing medical devices used for retinal acquisition and examination, followed by biometric devices. High-quality scans of the retina are provided by specialized medical devices. However, two significant disadvantages cause their failure in the biometric market. First is the very high cost, ranging from thousands (used devices) to tens of thousands of euro. Second, medical personnel is needed for data acquisition since medical devices only have the manual or semi-automatic mode. Up to now, there is still no device in the market that can work in fully-automatic mode (without user interaction). The device is still in development; however, its price is already estimated too high for the biometric market.

2.4.1 Medical devices

A frequently used ophthalmologist's device for examining the human retina is a *direct ophthalmoscope*. The doctor examines the retina through the pupil at only a few centimeters. The disadvantage of this device is the relatively small observed area and need cooperation of patients [11].

For a more thorough observation of the retina, it is appropriate to use a so-called *fundus camera* (as be seen at **Figure 3**). After a relatively long acquisition and focus, the camera takes a picture of the back of the eye in a short time and the doctor can go through it for any length of time, or look for changes in the retina since the last examination. By rotating the eye, a large area of the retinal area can be examined.

A *slit lamp* allows examining the eye's anterior segment using *biomicroscopy.* This, along with direct and indirect ophthalmoscopy, are the main ophthalmoscopic examination methods for the eye's anterior and posterior parts, where the *slit lamp* is the most widely used. A *fundus camera*, also known as a *retinal camera*, is a special device to display the optic nerve's posterior segment, the yellow spots, and the retina's peripheral part. It works by indirect ophthalmoscopy where the instrument has a built-in source of primary white light, which can be modified by different types of filters.

The optical system focuses on the human eye where the light reflects from the retina and bounces back to the lens of the fundus camera. If mydriasis has to be first applied to the eye, then a mydriatic fundus camera is used. Its intention is to enlarge the "inlet opening" of the pupil, which allows scanning a larger portion of the retina. Non-mydriatic fundus cameras are favored as no procedure is done that

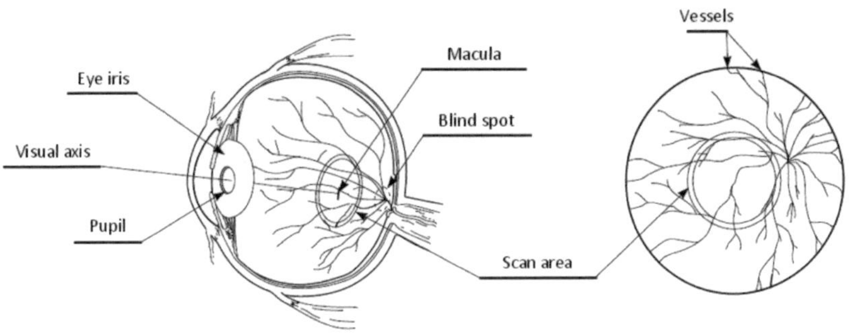

Figure 4.
Principle for obtaining an eye background image [12].

will alter the normal sight of the subject. Mydriasis, on the other hand, is required for some subjects. The costs of these devices are calculated by medically specialized workplaces and are in tens of thousands of euros.

The optical device has a complex mechanical construction. The scanning device works based on the concept of medical eye-optic devices. These retinoscopes or fundus cameras are complex devices which are also quite expensive.

The reflection of a part of the light that came from a beam and hit the retina is scanned by the CCD camera. This concept is similar to retinoscope, where the eye lens concentrates on the retina's surface due to the adjustment made to the beam of light that is coming from it. The ophthalmic lens receives back the reflection of only a part of the transmitted light beam and readjusts it. The beam leaves the eye below the angle where it entered the eye (return reflection). An image showing the eye's surface can be obtained at roughly 10° surrounding the visual axis, as in **Figure 4**. A circular snapshot of the retina is captured by the device from the reflection of light coming from the cornea, which would be useless in raster scanning [13].

3. Biometric recognition and verification

A biometric recognition system (BRS) recognizes a person by analyzing the random pattern in his/her physiological or behavioral characteristics known as biometrics, which are unique, non-transferable, unforgettable, and always carriable. Among biometric characteristics, we can also count voice, signature, face, gait, fingerprint, DNA, odor, vein pattern, hand geometry, signature, iris, retina, etc. A BRS has become a usual requirement in strictly protected areas such as nuclear plants, military facilities, scientific laboratories, cash vault, border, airport, government office etc., as well as in our day-to-day life such as online banking, car locking, building access, phone unlocking etc. Comparing to other biometrics, eye biometrics, which include iris and retina [8], offers higher degree of randomness. Even for identical twins the pattern of retinal blood vessels and iris are very distinctive [14, 15]. In addition, eye biometrics remain the same for the entire lifetime of a person. Therefore, the error rate of eye biometrics-based BRS is very low. Even though, both eye biometrics offer very high security, the probability of occurrence of counterfeiting is lower in the retina-based BRS (RBRS) than the iris based BRS. Because without users' cooperation and special camera like fundus camera or ophthalmoscope, it is not possible to capture retinal image. On the other hand, iris images can be captured by a normal camera at a distance.

In an RBRS, the unique pattern of blood vessels in the retinal image is used to recognize a person. Four kinds of approaches are generally found in the literature for capturing the uniqueness of retinal vessels, among which one approach is matching bifurcations and crossings of the blood vessel structure [16]. Bifurcation is the point where one blood vessel divides into two branches. The crossing is the point where blood vessels cross one another. Inspired by the idea of fingerprint minutiae [17, 18], Ortega et al., claimed in [9, 13] that using of bifurcations and crossings as feature points can overcome the drawback of the RBRS which uses the tree shape pattern of blood vessels of the whole retina proposed by Mariño et al. [19].

The retina is suitable for biometric purposes. As already mentioned, the pattern of blood vessels is unchanged during human life. In addition, the retina is well protected from the environment. However, this is also a disadvantage, because its capture is relatively complicated. In order to uniquely identify a person by capturing the uniqueness of the eye's tree-shaped blood vessels, four kinds of approaches have been proposed [16]: (i) using general signal and image processing techniques on the raw retinal images; (ii) matching of the branching blood vessel structure as a whole; (iii) matching bifurcations and crossings of the blood vessel structure; and (iv) matching the pattern of the vessels that are traversing a well-defined region. In all these approaches, a database is created by storing the templates made by the features in the training phase. These features are accessed in the identification phase.

The uniqueness or randomness of tree-shaped blood vessels can be measured by *biometric entropy* which has unit in bits. The bigger the biometric entropy, the lower the chance that two retinas of two different persons will match. There are two ways to determine biometric entropy [16]. The first one is to fit the distribution of the features to existing theoretical models and the second one is to determine empirically the probability p of matching templates of two different persons. In the second way, the entropy is $log_2 p$. Arkala et al., [16] measured the biometric entropy in the ring around the blind spot. Each vessel segment was represented by a triplet: *position* (in degrees around the ring), *width* (thickness in degrees), and angle (the angle that the segment makes with a radial line from the ring passing through the segment's centroid). The biometric entropy result was approximately 17 bits. That means 10^{17} possible combinations of retinal patterns.

3.1 History of retinal recognition

The image of the bloodstream in the retina was found to be unique for two individuals. This allows the discovery of eye diseases by ophthalmologists *Carleton Simon* and *Isidore Goldstein* in 1935. Later on, they released a journal article about identifying unique patterns in the retina using vein imaging [20]. Dr. Paul Tower supported it, and in 1955, released an article on the study of monozygotic twins [14]. The article states that the patterns of the retinal vessel have the least resemblance to all the other patterns that were examined. Identifying the retina of the vessels was an unchanging idea back then.

Robert Hill, the founder of EyeDentify in 1975, spent most of his time and effort in the development of a simple and fully automated device that can capture a snapshot of the retina and use it to verify the user's identity. These devices, however, did not emerge in the market even after several years [8, 21].

The remaining fundus cameras were modified by several other companies to capture an image of the retina to be used for identification. They, however, had many great disadvantages, such as the corresponding complex alignment of the optical axis, the visible light spectra, which causes discomfort, and the cameras being too expensive.

Infrared (IR) illumination was later discovered and used. Choroid reflects the radiation coming from the almost transparent beams that are hitting it. This reflection creates an image of the blood vessels in the eye. Since it is not visible, the pupil diameter is not reduced even when the eye is being irradiated.

The first prototype of the IR device was released in 1981. It has an eye-optic camera to illuminate the IR radiation. The camera was attached to an ordinary personal computer that will be used to analyze the captured image using a simple correlation comparison algorithm.

EyeDentification System 7.5 was launched after four years by EyeDentify Inc. Its verification is done using the retina image and the PIN entered by the user, with the user data stored in the database [8, 21].

ICAM 2001 was the last known retinal scanning device that was made by EyeDentify Inc. The device might have been able to store a maximum of 3,000 subjects with a storage capacity of 3,300 history transactions [8]. Unfortunately, the product was withdrawn due to low user acceptance and high price. Other companies, such as Retica Systems Inc., worked on a retinal acquisition device prototype for biometric purposes that might have been more user friendly and easier to integrate into commercial applications. Unfortunately, this device was also a failure in the market.

3.2 Limitations

Retinal biometrics limitations discourage further use of it as a biometric system. There are still no acceptable solutions found for these shortcomings [21].

Fear of eye damage - due to a myth about the devices damaging the retina. The level of infrared illumination used by these devices is low and has proven to be completely harmless. The people must be shared with this information so that they will not be afraid of using these devices.

Outdoor and indoor use - the return beam of the light passing through the pupil twice (once inward then outward of the eye) can be greatly weakened if the subject's pupil is too small. This can result to an increase in the false rejection rate.

Ergonomics - the subject must be near to the sensor, which may cause discomfort.

Severe astigmatism - the eye must be focused on a point. This may be difficult for those with visual impairments such as astigmatism, which can negatively affect the template generation.

High Price - the cost of optical devices is always more than the price of other biometric systems such as fingerprint or voice recognition capturing devices.

High-security areas such as nuclear and arms development, even manufacturing, government and military facilities, and other critical infrastructure can make use of retinal recognition.

3.3 Recognition schemes

Several schemes can be used for recognizing retinal images. For instance, a retina image biometric recognition has different approaches. Farzin [8] and Hill [21] have segmented the blood vessels to generate features and store, at maximum, 256 12-bit samples, which are then shrunk to a reference record containing 40 bytes for each of the eye. The Time-domain stores the contrast information. Fuhrmann and Uhl [22] extracted the vessels which obtained the retina code. The retina code is a binary code describing the vessels surrounding the optical disc.

3.4 Verification phase

In order to be able to use the proposed algorithm universally, and therefore also for the verification phase, it is necessary to choose the parameters with regard to the verification steps. During the verification phase, when recognizing samples that should be identical, we encounter the problem of inaccuracy in imaging. We practically never manage to take a picture in exactly the same way. There are small inaccuracies such as rotating or zooming on the image. These deviations must be ignored to a small extent.

Another problem may be the absence of some points. Even these inaccuracies can affect the similarity score obtained. In the verification phase, it must be with a relatively low penalty. The values in the previous chapter are therefore chosen so that the same algorithm can be used for both the recognition and verification phases.

4. Our recognition method

The distribution of vascular lines in the retina of the human eye is unique (as shown in Chapter 3.1), which is similar to the papillary line on the human fingers. Currently, there is no single approach to retinal recognition. Our procedure follows dactyloscopy, where bifurcations, terminations, positions and directions of a detected point are stored. We look for "anomalies" on the vessels in the retina - the places of visual crossings and bifurcations - and also record their position within the retina. For images, it is not easy to recognize whether it is a crossing or bifurcation of a vessel as the two phenomena often overlap. Therefore, we are only interested in the feature and not on its specific type. The termination of the vessel takes place "until lost" so a specific place cannot and will not be detected. We locate the points according to the position relative to the optical disk and the fovea. Therefore, we also store their position within the image as will be further described in Chapter 4.2 - the coordinate system. The result is a set of vectors such that the system is not affected from the changes in retinal scanning (different rotations, zooms, or chamfers).

Recognition becomes problematic in the presence of diseases that are manifested by a change in the retina such as bleeding. As with other biometric features, a relatively large amount of human health information can be read from retinal manifestations. Therefore, it is appropriate that the biometric facility manager guarantees the non-misuse or non-storage of this sensitive data, for example under the GDPR Directive [9, 23].

4.1 Statistical evaluation of the crossings and bifurcations frequency

If we take a brief look at a few images of the human eye retina, we discover that crosses and bifurcations are not equally frequented in various areas. The probability of their occurrence is in some areas higher, in others almost zero. In the beginning, it should be noted that the ability to mark intersections and bifurcations strongly depends on the quality and contrast of the image. In the statistically empty parts are the very small capillaries that are undetectable in the image using automatic or manual search.

When we create the frequency map, the points can be regarded with different weights to recognize the pattern. Finding matching points in two retinas being compared in rare occurring sites may score higher than matching points in other areas. Therefore, we tried to statistically evaluate several hundred retinal images and create our own frequency scheme, which we will later use to adjust the evaluation when comparing two retinas.

4.2 Coordinate system

In order to be able to work uniformly with all retinas without major complications, we have introduced a polar coordinate system, where two values can be used to align the retinal image to same coordinate system with the others. Our coordinate system assumes distances between the optical disk and the fovea in the different retinas approximately similarly. We also assume, if the physical structure of the retina differs significantly, that its development proceeded by similar rules. For example, if the distance between the optical disk and the fovea is smaller than average, the entire retinal structure will be smaller, and this will not affect our system.

The main point of the entire coordinate system is the center of the optical disk. In the records of individual retinas, its position in a particular image is stored as the distance from the left and top of the image edges. In addition, the width and height of the optical disk area (1st line of the output text file) are stored here as well. The second record is the center of the fovea (2nd line). The width and height of fovea are no longer stored here because its boundaries are difficult to ascertain by a simple look. The distance *r* between these two points is the basic unit of length for our coordinate system in each retina. This value may differ for every single image but is always valid for one retina. The second value is given point angle ψ of the direction from the optical disk. An angle of 0° lies in the line to the fovea and the value increases as it goes clockwise. This means that the fovea will have coordinates (1, 0°) in all retinas according to our coordinate system. Bifurcations or crossings are expressed by *r* and ψ, respectively, and are stored on the next lines of the output file.

We convert the found bifurcations and crossings back from the polar to the Cartesian coordinate system when we need to display or evaluate the entered points globally. To do this, we need the distance between the center of the blind spot (C_{BS}) and the yellow dot (C_{YS}). Then, we calculate their Euclidean distance (d) and the angle between the centers of both points (α) according to the Eq. (1).

$$\alpha = arctg2\left(\left(y \cdot C_{YS} - y \cdot C_{BS}\right), \left(x \cdot C_{YS} - x \cdot C_{BS}\right)\right) \tag{1}$$

Calculate the bifurcation/crossing distance from the blind spot using Eq. (2):

Using Eq. (2), we calculated the bifurcation/crossing distance from the blind spot:

$$v = r \cdot d \tag{2}$$

Then, calculate the coordinates *dx* and *dy* using Eqs. (3) and (4).

Then, using Eqs. (3) and (4), we calculated the coordinates *dx* and *dy*:

$$dx = d \cdot \cos(\psi + \alpha) \tag{3}$$

$$dy = d \cdot \sin(\psi + \alpha) \tag{4}$$

Lastly, calculate the point that resulted from the bifurcation/crossing in the Cartesian system $[dx + x \cdot C_{BS}; dy + y \cdot C_{BS}]$.

4.3 Recognition scheme

The algorithm for determining the grade of conformity of the two retinas works by converting all points from the polar coordinate system (described earlier) to Cartesian coordinate system. It is not necessary to align or rotate both images. Due to the chosen system, which is based on the position of the optical disk and the fovea, these points on two retinas will always exactly overlap.

For the first point of intersection and bifurcation in the first retina, it is determined which set of points in the second retina is the closest one. Then, the same procedure is repeated for the point found from second retina - finding the closest point back in first retina. This will determine if the two points are really the closest. If we did not find points on both directions, as in the case of a marked point in only one retina, another point could be found, which has its pair from the first retina.

Figure 5 shows the scenario without two-way control. Green points from the first retina and blue points from the second retina are combined into one image. For a point in a red circle, we look for the nearest point that belongs to a pair with another point.

If the distance between the two found points is less than the specified limit, then they are not considered as close. If they are "close", they will be removed from the lists of both retinas and their distance is saved. Before saving, the distance value is converted to a percentage where 0% means zero distance between points and 99% is the maximum allowed distance between two points to be considered as close. And this value is then squared for better difference between near and far points.

The percentage value of the distance is then adjusted according to the statistical model described above. The value is multiplied by a number from 1 to 4 where a lower number means lower frequency in the statistical model. The reason is, if no nearby point is found in the high frequency region, then it is much worse for retinal conformity than if no point is found between two far points that are at distances from the optical disk. In addition, places with higher frequencies are usually closer to the center of the coordinate system where it is more accurate.

4.4 Used databases

For the testing purposes, we used several public or our internal databases. Messidor [24], e-ophtha [10], High-Resolution Fundus (HRF) [25] and Retina EBD STRaDe (EBD).

First of them is publicly available database *Messidor* [24] from team ADCIS. The *Messidor* database has 1,200 eye fundus color numerical images of the posterior pole. The images were captured by three ophthalmologic departments by using a color video 3CCD camera on a Topcon TRC NW6 non-mydriatic retinograph with a field of view of 45°. The captured images use 8 bits per color plane at 440 × 960, 240 × 488, or 304 × 536 pixels. There are 800 images that were captured with the pupil dilated (one drop of Tropicamide at 0.5%) and 400 without it.

Second one is database *e-ophtha* also from team ADCIS. This database has 47 images that are with exudates and 35 images without lesions. For our purposes we use only images of healthy retinas.

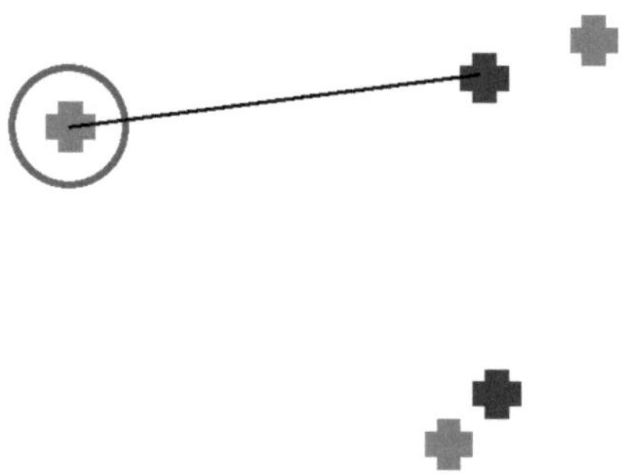

Figure 5.
Illustration of finding the nearest point.

Next is High-Resolution Fundus image database from German Fiedrich-Alexander University. The *HRF* database has 15 images of healthy patients, 15 images of patients with diabetic retinopathy, and 15 images of glaucomatous patients. Each image has a binary gold standard vessel segmentation image. Moreover, particular datasets are provided with masks to determine the field of view (FOV). A group of experts from the retinal image analysis field and the medical staff from the cooperating ophthalmology clinics generated the gold standard data.

The EBD is internal set of iris and retina images our research group STRaDe (Security Technology Research and Development at the Faculty of Information Technology, Brno University of Technology (CZ), focused on security in IT and biometric systems). The database contains 684 images of both retinas from 110 distinct people, totaling 220 distinct samples. Unfortunately, a significant part of this set consists of very low-quality pictures. But in this database all persons have several images of each eye.

For additional checking of our algorithms we use our retinal fundus camera at our laboratory, which we make for several past years. We use 30 images from students, which captured during Biometric systems course. Some images have bad quality, which is useful for testing applications in a worse condition. Several images are from same person eye (further as "school database").

4.5 Developed applications

We developed several application software modules to determine some properties of the retina, which will then be used to find out the degree of similarity of the two entered retina patterns.

4.5.1 Manual marking program

The first program (SW_1) is developed for manual retina marking by our students. First, the edges of the optical disk are marked. The program stores the top left position and the width and height of the ellipse around the optical disk. Then, the fovea is marked. Both positions are stored in Cartesian coordinates, which are based on the image properties and resolutions. Each feature is then marked after both main structures in the retina. These points are stored in polar coordinates. Data from the images are stored as a plain text file. By this program, we marked all retinal images using Messidor's e-optha and HRF databases.

4.5.2 Automatic marking program

The second program (SW_2) stores the same information about the image as SW_1, except that it performs the steps automatically. Details of the overall work of the program, its steps, and further development are summarized in work [5]. The program is developed in Python and was used on the same images as SW_1. An average of 48 features was found in these retinal images.

4.5.3 Compare program

SW_3 compares the detection accuracy between the manually marked-up results by SW_1 and the automatically marked-up results by SW_2. The algorithm is designed to compare the bifurcations/crossings that were selected manually, with the automatically detected set after they have been detected. The paired bifurcations/crossings are automatically found through a method like that in chapter 4.3.

The converted points are stored in the list with their position in it being used as ID for compiling disjunctive sets. A placeholder ID is assigned with −1 value. The problem is converted to an integer programming problem [26] to calculate the minimum pairing. The edges are then determined between the graphs' individual vertices. The number of bifurcations/crossings that were manually found and paired can then be calculated.

4.5.4 Visualization program

SW_4 collects the marked data by SW_1 and SW_2 in the previously described text file format into one picture. It collects individual pixels into a grid of adjustable size. For our purposes, a summary grid of 5 × 5 pixels was chosen as the most suitable. In the images, the fields with a higher frequency of occurrences are colored darker.

4.5.5 Recognition program

SW_5 works according to the recognition principle described in chapter 4.3. For this purpose, we used database EBD, which have more images from one retina. For practical tests we used our school database of retina images, which has taken by students of our faculty.

5. Results

The representative sample of data is obtained by randomly choosing 460 images, which came from Messidor, 160 images came from e-optha, and 50 images came from HRF. The chosen retinal images have both the left and right eye images available. They were shrunk to a resolution of around 1 Mpx to fit them on the screen. FIT BUT students in 2016 and 2017 did the manual marking of bifurcations and crossings by SW_1.

5.1 Accuracy of manual and automatic marking

After manually marking the points via SW_1, SW_2 also searched for the same points automatically. With this principle, we determine the accuracy of the positions of points during manual marking. The resulting average deviation of a minutiae was about 5 pixels [5].

Simultaneously, the automatic algorithm tested on the VARIA database [27], contained 233 images from 139 individuals, is enhanced. The resultant image emphasizing the comparison is shown in **Figure 6**.

On the contrary, we assume that the hand marking will accurately highlight the blind and yellow spots. SW3 also compares the correct location of the automatic search. The success rates were 92.97% for the blind spot and 94.05% for the yellow spot. Wrong localization of spots was mainly caused by too much brightness or darkness in it.

5.2 Frequency of features occurrences

Figure 7 shows the frequencies for manual marking by SW_1. The program SW_4 also shows the outer circle area in which at least some points occur, and the axes of the retina and the inner ellipses indicate areas with minimal occurrences of points around the yellow (yellow) and blind (black) spots.

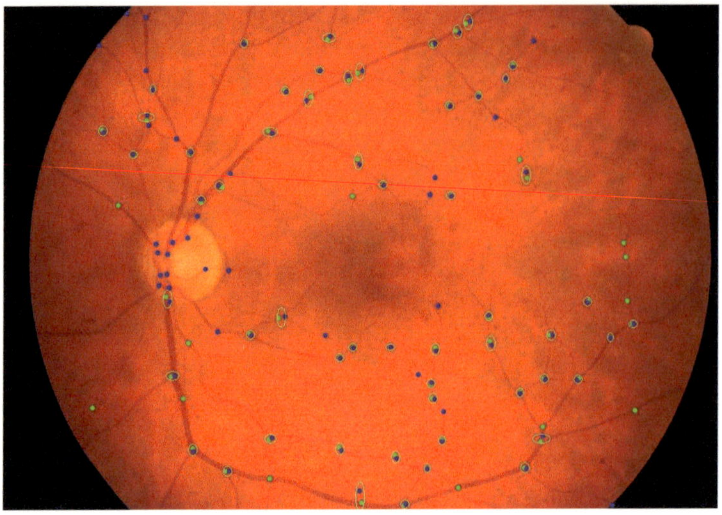

Figure 6.
Comparison of manual and automatic marking. Manually marked points are displayed by green points and automatically marked points are displayed by blue points.

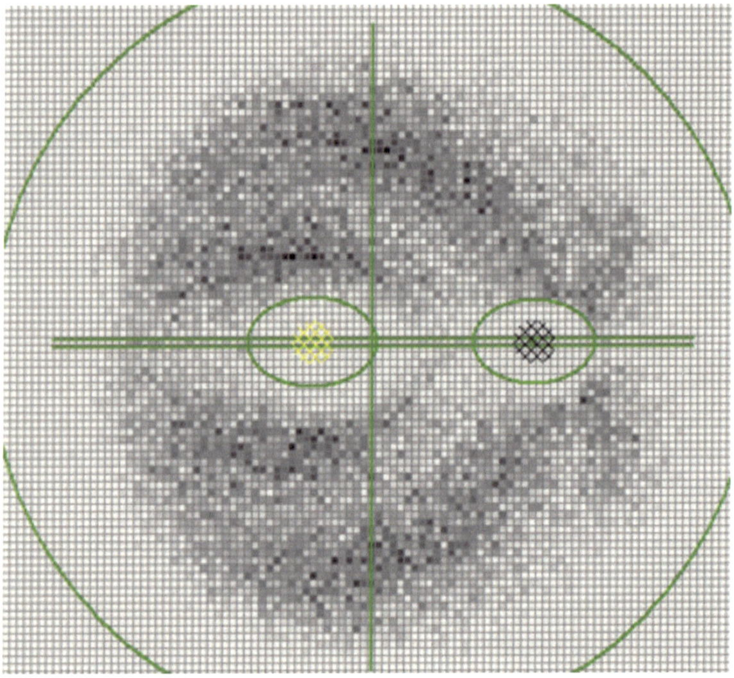

Figure 7.
Frequency visualization of bifurcations and crossings.

For SW_5, we randomly selected 10 persons from database EBD. Each have three images for the left and right eye which are marked e.g. 1002-2-R for person 1002's second image of the right eye. We mark all retinas by SW_1 very carefully. The first and third images of the eye were marked by a specialist such that at least 24 hours must have passed in between the marking of the retinas for the same person. The second images were marked by an ordinary computer user who is not involved in the project.

Tables 1 and **2** shows the average results of evaluation after SW_5.

	Avg value	Min value	Max value
compare with second eye of same person	65.52%	39.69%	80.38%
compare with different person eye	67.66%	45.32%	74.21%
compare with our school database (different persons)	57,62%	43,36%	74,58%

Table 1.
Recognition comparing results.

	Avg value	Min value	Max value
same eye, marked by different person	86.51%	78.90%	93.16%
same eye, marked by same person	93.54%	88.54%	95.78%

Table 2.
Verification results.

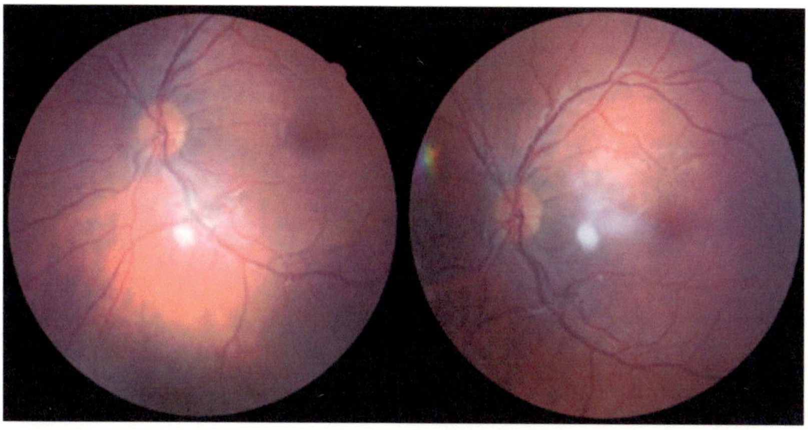

Figure 8.
Sample of two images of the same retina in the EBD database.

Figure 8 shows two images of the same retina that are taken at significantly different angles. After the marking by SW_1 and determining the similarity of the retinas by SW_5, the result was 91,09%.

6. Conclusions

In this paper, we have presented different approaches for the evaluation of individual parameters in human recognition based on the retina of the human eye. The main principle was to locate the individual bifurcations and crossings in the retinal image. The main part was, of course, based on a comparison of the locations of the points in both images. Another part of the principle was based on a set of almost 1000 manually marked images where all bifurcations and crossings were located. Were tested these points in images for placement accuracy by automatic search. Depending on the frequency of occurrence of points in different parts of the retina, these points had different weights for final correspondence. Finally, the procedure was tested on several differently photographed retinas of one person.

In our principle, it is not a matter of finding the best recognition algorithm but finding out the best properties of the individual parts of the procedure. The algorithms could be improved especially in the recognition part itself. The evaluation algorithm was illustrative only to show how the individual parts worked.

Acknowledgements

This work was supported by The Ministry of Education, Youth and Sports of the Czech Republic from the National Programme of Sustainability (NPU II); project IT4Innovations excellence in science - LQ1602 and "Secure and Reliable Computer Systems" IGA FIT-S-17-4014.

Author details

Lukáš Semerád* and Martin Drahanský
Faculty of Information Technology, Centre of Excellence IT4Innovations,
Brno University of Technology, Czech Republic

*Address all correspondence to: isemerad@fit.vutbr.cz

References

[1] Anatomy & Physiology, Connexions Web site. http://cnx.org/content/col11496/1.6/

[2] M. D. Jakobiec, A. Frederic, "Ocular Anatomy, Embryology and Teratology," Harpercollins, pp. 562-563, 1982.

[3] Barajas H. *Atlas of the Human Eye: Anatomy & Biometrics*. Palibrio, p. 74, ISBN 978-1506510330.

[4] Snell R.S., Lemp M.A. *Clinical Anatomy of the Eye*. 2nd Edition, Wiley-Blackwell, 2013, ISBN 978-0632043446.

[5] Pres M. *Bifurcation localization in retina images*. MSc. thesis, supervised by: Lukas Semerad, Brno University of Technology, Faculty of Information Technology, Czech Republic, 2016.

[6] Optimis Fusion: http://www.askin.cz/prednesegmentove/, accessed on 2018-06-06.

[7] Kowa VX-20: http://dfv.com.au/products/diagnostic/diagnostic-imaging/kowa-vx-20-mydriatic-non-mydriatic-integrated-fundus-camera/, accessed on 2018-06-06.

[8] Farzin H., Abrishami-Moghaddam H., Moin M.S. *A Novel Retinal Identification System*. EURASIP Journal of Advances in Signal Processing, Hindawi, Vol. 2008, p. 10, DOI 10.1155/2008/280635.

[9] K. Acharjya, G. Sahoo and S. Sharma, "An Extensive Review on Various Fundus Databases Use for Development of Computer-Aided Diabetic Retinopathy Screening Tool", Proceedings of ICSCSP 2018, 2019.

[10] Decencière E., Cazuguel G., Zhang X., Thibault G., Klein J.C., Meyer F., Marcotegui B., Quellec G., Lamard M., Danno R., Elie D., Massin P., Viktor Z., Erginay A., La B. and Chabouis A. *Teleophta: Machine learning and image processing methods for teleophthalmology*. IRBM, Elsevier Masson, Vol. 34, No. 2, pp. 196-203, 2013.

[11] Timberlake G.T., Kennedy M. *The Direct Ophthalmoscope – How it Works and How to Use it*. University of Kansas, 2005, p. 39, available online on: http://web.media.mit.edu/~raskar/Eye/TheDirectOphthalmoscope.pdf.

[12] Robert Hill. *Retina Identification*. Michigan State University. 2005

[13] E. Trucco, T. MacGillivray and Y. Xu, "Computational Retinal Image Analysis: Tools, Applications and Perspectives", Academic Press, 2019.

[14] Tower P. *The Fundus Oculi in Monozygotic Twins: Report of Six Pairs of Identical Twins*. AMA Arch Ophthalmology, No. 54, pp. 225-239, 1955.

[15] J. Daugman, "How iris recognition works," IEEE Transaction on Circuits and Systems for Video Technology, vol. 14, no. 1, pp. 21– 30, 2004.

[16] Arakala, A., Culpepper, J. S., Jeffers, J., Turpin, A., Boztas, S., Horadami, K. J., McKendrick, A. M. Entropy of the Retina Template. Berlin, Heidelberg: Springer Berlin Heidelberg, 2009, pp. 1250-1259.

[17] A. Jain, L. Hong, S. Pankanti, and R. Bolle, "An identity authentication system using fingerprints," Proceedings of the IEEE, vol. 85, no. 9, pp. 2127-2134, 1997.

[18] X. Tan and B. Bhanu, "A robust two step approach for fingerprint identification," Pattern Recognition Letters, vol. 24, pp. 2127– 2134, 2003.

[19] C. Mariño, M. G. Penedo, M. J. Carreira, and F. González, Retinal

Angiography Based Authentication. Berlin, Heidelberg: Springer Berlin Heidelberg, 2003, pp. 306-313.

[20] Goldstein I., Simon C. *A New Scientific Method of Identification*. New York State Journal of Medicine, Vol. 35, 1935.

[21] Hill R.B. *Retina Identification*. In: Biometrics: Personal Identification in Networked Society. New York: Springer, 1996, pp. 123-141.

[22] Fuhrmann T., Hämmerle-Uhl J., Uhl A. *Usefulness of Retina Codes in Biometrics*. Advances in Image and Video Technology, 3rd Pacific Rim Symposium, Japan, 2009, Springer LNCS 5414, ISSN 0302-9743.

[23] J. Hájek and M. Drahanský, "Biometric-Based Physical and Cybersecurity Systems: Recognition-Based on Eye Biometrics: Iris and Retina", Springer International Publishing, 2019, p. 37-102.

[24] Decencière E., Zhang X., Cazuguel G., Lay B., Cochener B., Trone C., Gain P., Ordonez R., Massin P., Erginay A., Charton B. and Klein J.C. *Feedback on a publicly distributed database: the messidor database*. Image Analysis & Stereology, Vol. 33, No. 3, pp. 231-234, 2014.

[25] Kohler T., Budai A., Kraus M., Odstrčilík J., Michelson G. and Hornegger J. *Automatic no-reference quality assessment for retinal fundus images using vessel segmentation*. 26th IEEE International Symposium on Computer-Based Medical Systems, pp. 95-100, 2013.

[26] Shih H.P. *Two Algorithms for Maximum and Minimum Weighted Bipartite Matching*. Department of Computer Science and Information Engineering National Taiwan University, 2008.

[27] M. Ortega, M. G. Penedo, J. Rouco, N. Barreira, M. J. Carreira, "Retinal ration using a feature points based biometric pattern", EURASIP Journal on Advances in Signal Processing, vol. 2009, Article ID 235746, 13 pp., 2009.

Chapter 6

New Attributes Extraction System for Arabic Autograph as Genuine and Forged through a Classification Techniques

Anwar Yahya Ebrahim and Hoshang Kolivand

Abstract

The authentication of writers, handwritten autograph is widely realized throughout the world, the thorough check of the autograph is important before going to the outcome about the signer. The Arabic autograph has unique characteristics; it includes lines, and overlapping. It will be more difficult to realize higher achievement accuracy. This project attention the above difficulty by achieved selected best characteristics of Arabic autograph authentication, characterized by the number of attributes representing for each autograph. Where the objective is to differentiate if an obtain autograph is genuine, or a forgery. The planned method is based on Discrete Cosine Transform (DCT) to extract feature, then Spars Principal Component Analysis (SPCA) to selection significant attributes for Arabic autograph handwritten recognition to aid the authentication step. Finally, decision tree classifier was achieved for signature authentication. The suggested method DCT with SPCA achieves good outcomes for Arabic autograph dataset when we have verified on various techniques.

Keywords: Arabic autograph verification, adaptive window positioning, (DCT + SPCA) method, feature selection, classification techniques

1. Introduction

Handwritten autograph plays an important role in modern life as it is routinely used in every sphere of human activity. Couto [1] utilizes a lexical similarity technique for each entity identified. This frequently makes it unattainable to differentiate between a forged signature and a signature created under influence. Chung [2] applied Fuzzy groups to handle uncertainty. Although there are contributing studies in this area, research often failed to take into account the influence of contributing factors such as distractions and singers' stress which may affect the signatures being signed [3, 4]. It is widely used for authenticating financial and business transactions [5, 6]. There are online and offline authentication systems. In contrast, online signature systems require special hardware such as pressure tablets. These devices extract dynamic information including pressure, signer's speed, and the static image of signature. Unfortunately, both online and offline signatures can easily be imitated or forged, leading to false representation or fraud [7]. Yang [8] used learned dictionary to check samples. This method has been successfully utilized in image recognition lately. According to Alattas [9], financial institutions are interested to benefit from the reliability and safety of offline signature-recognition systems. Another major reason is that online authentication systems require more complex processing and high-tech gadgets than off-line systems. Offline autographs are usually presented on a piece of paper, which is the norm in documentation. Currently, there is a need for efficient online and offline systems to ascertain the genuineness of personal autographs. Authentication of handwritten autographs usually consists of a series of procedures. These processes are pre-processing (where images are enhanced, binarized, divided into fragments and other related operations), feature extraction (features of the signatures are extracted as raw forms), feature selection or reduction (extracted features are reduced for efficiency), identification and authentication of the signatures against the signature database based on the selected features. A good verification outcome can be performed by likening the strong features of the taster against the autograph of a signer sample utilizing suitable techniques or classifiers [10]. Methods depend on local tests, which concentrate on the analysis of the essential features of different scripts [10–12]. Some studies utilized evolving curves which do not move away to near by features decreasing the superfluous fragmentation [13]. Based on the available gap in the literature, in this paper, we propose a new process to identify and authenticate Offline-Arabic signatures. This method uses a combination of techniques including adaptive window positioning procedure for autograph attribute extraction and feature selection method for reduced features and selection of important features. In this paper, enhanced Discrete Cosine Transform (DCT) and, Spars Principal Component Analysis (SPCA) method is used to extract attributes. Further, these extracted features are reduced to the best features only. In this research, in order to classify genuine and forged signature two types of classifiers: 1) Decision Tree and 2) Support Vector Machine (SVM) are applied. The classification outcomes of Decision Tree and SVM are compared to choose a better classifier.

2. Proposed scheme

In this part, we introduce an offline Arabic signature identification system based on classification techniques. The procedure consists of four phases: pre-processing, features extracting, selected feature by (DCT+ SPCA) technique, and matching. The complete process begins with acquiring the images of signatures to undergo a pre-processing stage, and then identification and verification process, which are illustrated in **Figure 1**.

2.1 Pre-processing

In this step, data are acquired and signature images are pre-processed. For the purpose of this study, Arabic signature is used as the data consisting of 500 true samples and 250 forged samples was used. True samples were obtained from 50 different persons. Every signer was asked to sign 10 times using common types of pens. The 10 signatures collected from each person were used as follows: six of these signatures were selected at random for system learning and the remaining four were used for system testing in addition to "ve forged" samples.

2.1.1 Arabic signature database

This study employed the Arabic signature database created by "Anwar Yahya Ebrahim" as the Arabic signature samples for testing the proposed method. The

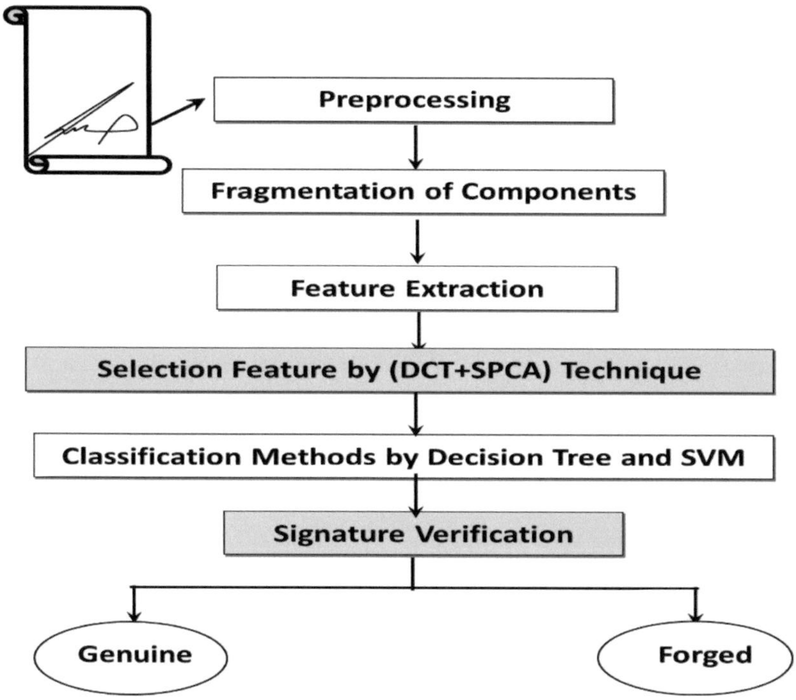

Figure 1.
Proposed methodology.

Arabic signatures are booked on A4 size paper and then scanned at 300 dpi, 256 gray level images. The dataset contains encompasses signatures from persons scanned signatures were collected from the signer Anwar Yahya Ebrahim et al. Each signatory has 10 signatures to predict a response of which, 6 are genuine signatures and 4 are forged signatures. There are enough signatures to ensure sufficient samples for both training and testing. Where 7 of the samples are assigned to the training set, and the rest 3 to the testing set from both classes.

The distribution of the number of genuine and forgery samples for different signatories is illustrated in **Figure 2**. Arabic Signature images are then pre-processed in order to improve the quality of images. Noises, such as irrelevant data, are removed from the features to develop the achievement of identification. These samples are then converted into binary samples before feature extraction process [14–17].

2.2 Feature extraction

Adaptive window positioning technique is then applied to separate Arabic autograph images into small segments or sub-images. This makes the process of removing redundant data easy and facilitates the comparison of segmented fragments. A 14x14 segment size is chosen for the images for an optimum output [18]. Further, a group of features (form measures) from the approaches are extracted, which represents the signature image in a feature space. To analysis data accurately, a variety of observations as well as a value of significant individual features are needed to be organized. Such the data can be given and analyzed by machines or humans.

The goal of form representation is to get form measures. These measures are used as classification features in models. Moreover, sub-images are presented from the set of obtained features [19].

Figure 2.
Examples of genuine signatures and their respective forged counterparts found in the Arabic signatures.

The attributes are then normalized using a feature matrix. The normalization process is very important. This is because when attributes are in various ranges, higher ratios may dominate lower values, which may distort the results. Normalization places the attribute ratios within the same scales and ranges to enable comparison. The projection and profile features are normalized by using window height, while the other descriptors are normalized by their maximum possible respective values. After normalization, each feature of the main window is composed to form a vector. This scales and translates each feature individually to a fixed range on the training set, which is a number between zero and one [20].

2.3 Features selection

This study proposes two fusions of features namely, Discrete Cosine Transform⊕ Spars Principal Component Analysis and (DCT⊕SPCA). The former is introduced represent the high pass in vertical, diagonal and horizontal directions, respectively in signature images whereas the latter is proposed to discriminate between genuine and forged of Arabic signatures. The reason to combine DCT and SPCA features is that both are transformed based features so due to homogeneity they are best choice for combining. Fusion combines the useful information from both images. The motivation to combine these both features are numerous similarities found in DCT and SPCA features. This proposed technique uses the high pass signature images to extract the necessary information for the signature verification.

Succeeding the feature selection, the twelve DCT features and the eight SPCA features are extracted. These features are then fused in order to classify signatures into genuine and forged classes. Suppose twelve DCT features are represented by

DOI: http://dx.doi.org/10.5772/intechopen.96561

$\alpha 1, \alpha 2, \alpha 3, \dots \dots \dots, \alpha 12$ and eight SPCA features are represented by $\beta SPCA1, \beta SPCA2, \beta SPCA3, \dots \dots, \beta SPCA8$. These both subsets of features can be combined by concatenating DCT features with SPCA features to form a single features vector (DCT⊕SPCA) of 20 features as shown below in Eq. (1).

$$\mathrm{DCT} = [\alpha_1, \alpha_2, \alpha_3, \alpha_4, \alpha_5, \alpha_6, \alpha_7, \alpha_8, \alpha_9, \alpha_{10}, \alpha_{11}, \alpha_{12}]$$

and

$$\mathrm{SPCA} = [\beta\mathrm{SPCA1}, \beta\mathrm{SPCA2}, \beta\mathrm{SPCA3}, \beta\mathrm{SPCA4}, \beta\mathrm{SPCA5}, \beta\mathrm{SPCA6}, \beta\mathrm{SPCA7}, \beta\mathrm{SPCA8}]$$

$$(\mathrm{DCT} \oplus \mathrm{SPCA}) = [\alpha_1, \alpha_2, \alpha_3, \alpha_4, \alpha_5, \alpha_6, \alpha_7, \alpha_8, \alpha_9, \alpha_{10}, \alpha_{11}, \alpha_{12}, \beta\mathrm{SPCA1}, \beta\mathrm{SPCA2}, \beta\mathrm{SPCA3}, \beta\mathrm{SPCA4}, \beta\mathrm{SPCA5}, \beta\mathrm{SPCA6}, \beta\mathrm{SPCA7}, \beta\mathrm{SPCA8}] \quad (1)$$

This set of 20 features represents one signature.

2.4 Classification

In this step, the model is presented based on training and testing. The various performed sub-steps are as follows:

3. Signature alignment

In order to perform a meaningful comparison of images of different lengths, we applied Extreme Points Warping (EPW) method [21]. EPW method modifies a shape using peaks and valleys as pivoting points, rather than warping the whole shape. The algorithm fixed the optimum linear alignment of two vectors by using the smallest overall dimension between them. The distances were recalculated between feature directions at each iteration. The alignment was considered to achieve optimal status in case the average dimension between feature vectors attained a low value. The dimension between two signature samples was calculated as the median of the dimensions between the fully aligned feature vectors.

3.1 Enrolment

For enrolment to the system, 54 signatures were selected from each user for training. Each pair of Arabic signatures was aligned to determine their distance, as described in the previous section. Using these aligned distances, the following measurement were evaluated:

1. Median dimension to the farthest sample (dmax).

2. Median dimension to the nearest sample (dmin).

The training group of Arabic signature images was used to determine the threshold parameter in order to distinguish dubious group from the genuine class.

4. Training

The 2-dimensional feature vectors (Pmin, Pmax) and normalize the feature values by the matching averages of the reference set (dmin, dmax) were obtained using the EPW algorithm. These were calculated based on Eqs. (2) and (3) to represent the allocation of the feature group.

$$N\ max = dmax/Pmax \tag{2}$$

$$N\ min = dmin/Pmin \tag{3}$$

Normalization of information ensures the genuineness or forgery of signatures in the training set. We trained a decision tree classifier to recognize the genuine and forged signatures in this normalized feature area (**Figure 3**). To facilitate comparisons, two classifiers were used: The Tree classifier and SVM classifier were applying the 2-dimensional attribute vectors. A linear classification was made by choosing a threshold ratio separating the two classes within the training set. This threshold was used in the verification process.

4.1 Classification based on SVM

For offline Arabic signature verification and identification Support Vector Machines (SVM) was used. Important features in the Arabic signature images were extracted and the samples were confirmed with the assistance of Gaussian empirical law. SVM was applied to record corresponding results for comparing all signatures from database with the test signature. The suggested method is tested on Arabic signatures containing 500 samples of 50 users and the outcomes are obtained to be encouraging. In a high dimension feature area the principle of SVM, depends on a linear isolation where information were mapped to take into consideration the final non-linearity of the issue. SVM classifier [22, 23] was trained with corresponding

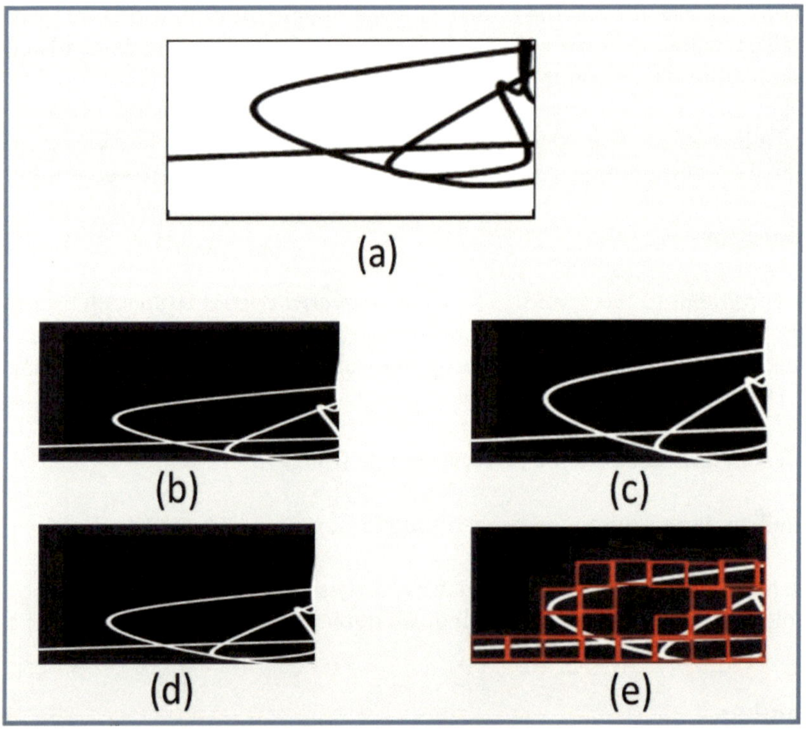

Figure 3.
The different stages of the pre-processing phase, (a) gray sample, (b) binary and converted sample, (c) with boundary box sample, (d) resized image, (e) windowing image.

result vectors for each distance. This is to obtain a good level of generalization capability. To establish the rating of signers' relationship to the inquiry samples, firstly we used these processing points and then we combined the results of the entire samples.

4.2 Decision tree classification

Evaluation of Tree Classification (Bagged Trees) technique was used in the same way and on the same samples from Arabic signatures as SVM. MATLAB 2014 bagged tree classification and trees software were used in the training and classification simulation. To predict a reaction, the decision procedure in the decision tree from the root (starting) node (feature) down to a leaf (feature) node was followed. Responses were included in the leaf feature. Decision trees granted responses, such as 'true' or 'false'. Decision Tree was created to perform classification [20, 24]. The described steps are presented in Algorithm 1.

Algorithm 1

Step 1: Start the first with all input features and then examine all potential binary divides on each predictor
Step 2: choose a divides with good optimization standard
Step 3: If the divide leads to a child node with less than the least leaf parameter), choose a divide with the better optimization standard. Subject to the least feature constraint
Step 4: put the divides and reiterate recursively for the two child (features) nodes
Step 5: If it made up of only observations of one category a (feature) node is perspicuous. Therefore, the node is fewer than minimum parent observations

5. Outcomes and discussion

In this section, we discuss the outcomes of the suggested methodology on some of samples from the Arabic signatures.

5.1 Pre-processing

The input image in RGB color space was first converted to grayscale image as displayed in **Figure 3(a)** represented Gray image. Then, the image was smoothened with median filter and converted to binary as shown in **Figure 3(b)**. Further, the image was passed from boundary box to find the boundaries of the text area as presented in (c), while in (d) the image was resized to apply the adaptive windowing algorithm to divide it into fragments as shown in (e).

5.2 Feature extraction

In this phase, we represent the sub-images from a set of features. The outcome of the feature extraction is shown in **Table 1(a)**. Initially, these features were not normalized. The values shown in **Table 1(a)** represent the frequencies of the designs extracted from each box. Higher ratios mean there is a more specific model with the genuine autograph, which suggests that the Arabic signatures are highly similar to the test signature. The features were then normalized using a composed matrix of features. The projection and profile features were normalized using window height, while the other descriptors were normalized by their respective maximum possible value. Normalization places different feature values in the same ranges as shown in **Table 1(b)**. After normalization, each normalized feature of main window were concatenated into a single feature set, which represent each

(a) Un-normalized features									
F_{10}	**F_9**	**F_8**	**F_7**	**F_6**	**F_5**	**F_4**	**F_3**	**F_2**	**F_1**
3.000	2.0020	3.0000	1.0001	3.000	1.0000	1.0000	6.0000	1.0001	3.000
1.0000	3.0000	1.0000	1.0000	1.0000	1.0000	1.0000	8.0000	1.0000	1.0000
1.0000	4.0000	1.0000	1.0000	1.0000	1.0000	1.0000	9.0000	1.0000	1.0000
1.0000	3.0000	1.0000	1.0000	1.0000	1.0000	1.0000	11.0000	1.0000	1.0000
1.0000	6.0000	1.0000	1.0000	1.0000	1.0000	1.0000	1.20000	1.0000	1.0000
1.0000	8.0000	1.0000	2.0000	1.0000	2.0000	2.0000	10.0000	2.0000	1.0000
2.0000	2.6463	2.0000	3.0000	2.0000	2.6463	2.6463	10.0000	3.0000	2.0000
3.0000	3.9281	3.0000	1.0000	3.0000	3.9281	3.9281	9.0000	1.0000	3.0000
4.0000	2.491	4.0000	1.0000	4.0000	2.491	2.491	6.0000	1.0000	4.0000
3.0000	1.8671	3.0000	1.0000	3.0000	1.8671	1.8671	3.0000	1.0000	3.0000
6.0000	1.3205	6.0000	1.0000	6.0000	1.3205	1.3205	1.0000	1.0000	6.0000
8.0000	2.6463	8.0000	1.0000	8.0000	0.3367	0.3367	1.0000	1.0000	8.0000
1.0000	0.9079	9.0080	1.3205	1.0000	0.836	0.839	1.0000	1.0000	9.0900
F20	**F19**	**F18**	**F17**	**F16**	**F15**	**F14**	**F13**	**F12**	**F11**
3.0000	2.0020	3.0000	4.088	2.6106	3.0000	1.0001	3.000	1.0000	6.0000
1.0000	3.0000	1.0000	4.764	1.057	3.0000	1.0000	1.0000	1.0000	8.0000
1.0000	4.0000	1.0000	4.472	1.5523	3.0000	1.0000	1.0000	1.0000	9.0000
1.0000	3.0000	1.0000	7.352	0.0523	1.0000	1.0000	1.0000	1.0000	11.0000
1.0000	6.0000	1.0000	5.336	0.1469	1.0000	1.0000	1.0000	1.0000	1.20000
1.0000	8.0000	1.0000	3.152	1.6021	2.9066	2.0000	1.0000	2.0000	10.0000
2.0000	2.6463	2.0000	3.376	1.0000	1.6974	3.0000	2.0000	2.6463	10.0000
3.0000	3.9281	3.0000	6.424	1.0000	1.0000	1.0000	3.0000	3.9281	9.0000
4.0000	2.491	4.0000	2.4	1.0000	5.0000	1.0000	4.0000	2.491	6.0000
3.0000	1.8671	3.0000	0.024	1.0000	4.0000	1.0000	3.0000	1.8671	3.0000
6.0000	1.3205	6.0000	0.304	1.0000	5.0000	1.0000	6.0000	1.3205	1.0000
8.0000	2.6463	8.0000	0.056	2.0000	6.0000	1.0000	8.0000	0.3367	1.0000
9.0080	1.0200	3.0010	0.464	1.3205	1.0000	1.3205	1.0000	0.836	1.0200
(b) Normalization									
F10	**F9**	**F8**	**F7**	**F6**	**F5**	**F4**	**F3**	**F2**	**F1**
0.690	0.57	0.824	0.635	0.041	0.61	0.68	0.72	0.83	0.65
0.087	0.345	0.760	0.622	0.706	0.157	0.156	0.395	0.760	0.67
0.523	0.875	0.8887	0.2794	0.472	0.523	0.0921	0.875	0.8887	0.2794
0.0523	0.477	0.3109	0.6446	0.352	0.0523	0.4585	0.477	0.3109	0.6446
0.1468	0.322	0.5577	0.424	0.396	0.1468	0.1098	0.322	0.5577	0.424
0.6021	0.8581	0.9066	0.6012	0.152	0.6021	0.0582	0.8581	0.9066	0.6012
0.2531	0.6463	0.6974	0.6831	0.376	0.2531	0.4802	0.6463	0.6974	0.6831
0.3451	0.9281	0.7784	0.1576	0.424	0.3451	0.2093	0.9281	0.7784	0.1576
0.6649	0.491	0.9262	0.0621	0.411	0.6649	0.6716	0.491	0.9262	0.0621
0.8189	0.8671	0.9862	0.4585	0.024	0.8189	0.1161	0.8671	0.9862	0.4585

(b) Normalization

F10	F9	F8	F7	F6	F5	F4	F3	F2	F1
0.6633	0.3205	0.9257	0.1098	0.304	0.6633	0.5974	0.3205	0.9257	0.1098
0.794	0.3367	0.9165	0.0582	0.056	0.794	0.4185	0.3367	0.9165	0.0582
0.6213	0.8938	0.9129	0.4802	0.199	0.6213	0.0595	0.8938	0.9129	0.4802
F20	**F19**	**F18**	**F17**	**F16**	**F15**	**F14**	**F13**	**F12**	**F11**
0.299	0.611	0.547	0.725	0.581	0.910	0.299	0.611	0.725	0.581
0.6727	0.741	0.0484	0.62	0.76	0.0157	0.6727	0.741	0.62	0.76
0.386	0.705	0.348	0.2794	0.472	0.523	0.386	0.705	0.2794	0.472
0.8651	0.925	0.6883	0.6446	0.352	0.0523	0.8651	0.925	0.6446	0.352
0.952	0.274	0.964	0.424	0.398	0.1568	0.952	0.274	0.424	0.398
0.4175	0.645	0.2759	0.612	0.152	0.6021	0.4175	0.645	0.612	0.152
0.915	0.722	0.7266	0.6831	0.376	0.2531	0.915	0.722	0.6831	0.376
0.9235	0.596	0.794	0.1576	0.424	0.3471	0.9235	0.596	0.1576	0.424
0.4185	0.666	0.9817	0.021	0.411	0.6649	0.4185	0.666	0.021	0.411
0.1315	0.231	0.9571	0.4585	0.024	0.8789	0.1315	0.231	0.4585	0.024
0.3969	0.666	0.4075	0.1098	0.304	0.6673	0.3969	0.666	0.1098	0.304
0.2144	0.754	0.8988	0.0582	0.056	0.7794	0.2144	0.754	0.0582	0.056
0.479	0.461	0.016	0.482	0.199	0.6613	0.479	0.461	0.482	0.199

Table 1.
Feature extraction un-normalized and normalized. (a) Un-normalized features (b)Normalization.

window by a vector. This process can standardize all features by scaling each feature to a given range.

5.3 Representation of feature selection

When the procedure of feature selection technique for windows was accomplished, those features with sufficient number of windows were kept. The features contained stroke patterns occurring in the windows. Generally, the number of patterns for each feature selection was proportional to the size of the Arabic signature sample. According to **Figure 4**, one important point to note is the number of selected features. This is a property of the signer as can be observed from **Figure 5**, where the number of selected features are presented. In this case, feature selection

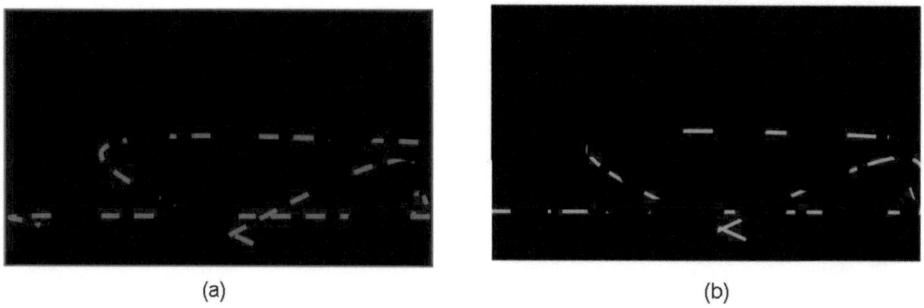

Figure 4.
After Selection Feature step by SPCA method.

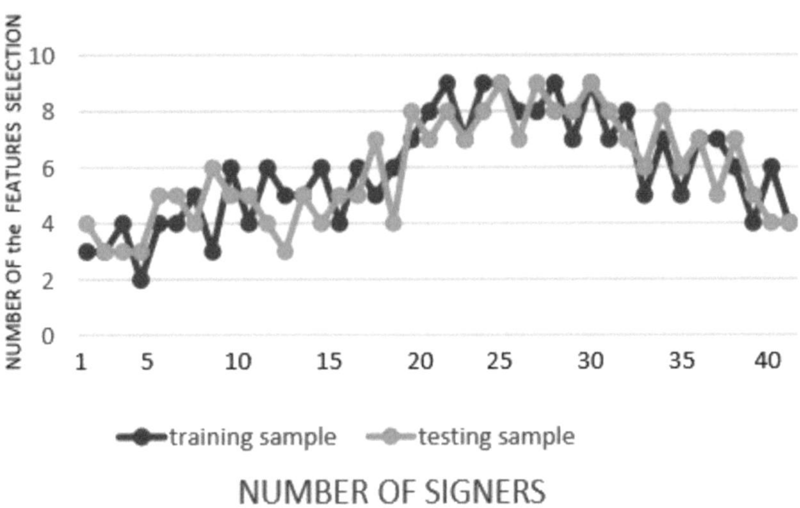

Figure 5.
The number of selected important features of DCT + SPCA method for the two samples of 40 signers.

is generated from 40 different signers using two tasters from each one. As can be realized, the bows represent the number of selected features in the two tasters of the same writer are close to each other for DCT + SPCA method. This seems consistent with the supposition that the value of selected features is a signer-dependent feature.

5.4 Matching

The matching phase is when the model is created using Classification and Regression Tree (Tree) and Support Vector Machines Classification (SVM) with different input parameters. Based on a person's signature, a model was created for the original and forgery signatures. The performance of the proposed method on 100 signers from Arabic signatures were used in identification for classification using DCT + SPCA features for selected important features with SVM classifier achieved the verification rate of 98.7%, and EER of 1.90% and same DCT + SPCA features with Tree classification achieved the verification rate of 99.8%, and the EER dropped to 1.20%. which was better than other techniques, as shown in **Table 2**. The objective of this study was creating a system that 1) can identify handwritten signatures and verify their authenticity, and 2 distinguish forgery from genuine ones, and those created under pressure and other influences. Using 2000 Arabic signatures samples. The results of the matching phase are shown in **Table 2**.

This implies that a forger may not skillfully repeat all aspects of the original signature. It also shows a pattern in forgers, which has small variations. Evidence shows that the mean of a feature produced by a forger in multiple attempts at forging tends to lie in a small range. Conversely, genuine signatures produced by a signer may vary under unusual conditions. Signers possess certain unconscious

Classification Techniques with Features Selection Technique	Verification Rate	Verification EER	Recognition Rate
Tree+ DCT+SPCA method	99.8%	1.20	98.5%
SVM+ DCT+SPCA method	98.7%	1.90	97%

Table 2.
Experimental results obtained from 100 signer based on Arabic signatures.

features that remain consistent and stable despite the interference of influencing factors. Such natural features are almost impossible to imitate, even by the original signers.

6. Authentication of results

The comparison between Arabic signature recognition methods were by verification rate and not by the computational time. The accuracy performance measure has been computed using confusion matrix Where, TP signifies the number of true positive signatures, TN refers to the number of true negative signatures, FP signifies the number of false positive cases and FN signifies the number of false negative signatures. True Positive Ratio is the measure of genuine signatures classified correctly as genuine; False Positive Rate is the measure of a forgery signature classified as genuine. False Negative Rate is the measure of a genuine situation classified as forgery. True Negative Proportion is a measure of a forgery signature classified as forgery.

$$\text{The Verification Rate} = (\text{TP} + \text{TN})/(\text{TP} + \text{TN} \ + \text{FP} + \text{FN}) \times 100\%$$

The tests assumed that 99.8% accuracy proportion Predicted Valu and Decision Tree. Such promising results are pinpointing the state-of-the-art preprocessing techniques and best performance of proposed features to discriminate between genuine and forged signatures with higher accuracy rate. The authentication of the achievements of the suggested method was achieved applying the verification rate using DCT + SPCA method were computed and compared against the two other vastly agreeable autograph verification methods. **Table 3** shows the simulation results with the Arabic signatures consisting 2000 signatures from 100 various signers. The validation rate for the proposed technique is 99.8% attesting to its superiority against the others. We could conclude that DCT + SPCA features technique and Decision Tree classifier was a credible and reliable technique for verification of offline Arabic signatures.

Authors	Methods	Language	Verification Rate
Ismail al. [25]	New procedures for autograph verification by fuzzy concepts	Arabic	98%
A.y. Ebrahim [26]	DCT+ DWT Technique	Arabic	99.75%
SM Darwish, al., [27]	Distance and Fuzzy Classifiers Alliance	Arabic	98%
C. Ergun al., [28]	word layout signature	Farsi	94:3%
Proposed method (2021)	DCT+SPCA features Technique	Arabic	99.8%

Table 3.
An assessment table relating between the projected Arabic signature recognition system based on Arabic signatures and other signatures with other previously known approaches.

7. Conclusion

This paper, we described a method we developed to important features selection using DCT + SPCA features technique in offline Arabic signature verification. It employed the partition of signature samples into 14x14 windows and generated the features extracted for each window. Then, this feature selection was used for classification techniques.

We have mentioned the limitation of the research in the apply of set of Arabic signatures for collecting the Arabic signature samples used in this study. To judge our findings objectively, we used Arabic signatures, which includes Arabic signers. The results of our study show that this method was a credible technique for offline Arabic signature feature selection. This method can be used as a Arabic signature verification method for the exposure of offline signatures.

In the simulation phase, two different comparisons have been made. The first was the performance of support Vector Machine classifier and DCT+ SPCA features technique, and the second was the performance of Decision Tree classifiers with DCT+ SPCA features technique working together. The Decision Tree classifiers and DCT+ SPCA features technique produced the best verification rate of 99%, which improved the performance of offline Arabic signature verification.

There are many extensions which can be employed to develop the study. The proposed future works can be divided into two main fragments. Firstly, the extension is by an expansion of the procedures with more accuracy for autographs verification. Secondly, the extensions which can be made to the different dataset of autographs.

Author details

Anwar Yahya Ebrahim[1,2]* and Hoshang Kolivand[1,2]

1 Department of Computer Science, Babylon University, Babylon, Iraq

2 Department of Computer Science, Liverpool John Moores University, Liverpool, UK

*Address all correspondence to: anwaralawady@gmail.com

References

[1] Couto, F. M., Silva, M. J., & Coutinho, P. M. (2005) Finding genomic ontology terms in text using evidence content. BMC bioinformatics, 6(Suppl 1), S21.

[2] Chung, F. L., Deng, Z., & Wang, S. (2009) An adaptive fuzzy-inference-rule-based flexible model for automatic elastic image registration. Fuzzy Systems, IEEE Transactions on, 17(5), 995–1010.

[3] Ben Jlaiel, M., et al. (2007) Three decision levels strategy for Arabic and Latin texts differentiation in printed and handwritten natures. Document Analysis and Recognition, 2007. ICDAR 2007. Ninth International Conference on. Vol. 2. IEEE (2007).

[4] Shrivastava, Shailedra Kumar, and Sanjay S. (2010) Gharde Support vector machine for handwritten Devanagari numeral recognition. International journal of computer applications 7.11: 9–14.

[5] Arora, Sandhya, et al. (2010) Performance comparison of SVM and ANN for handwritten devnagari character recognition. arXiv preprint arXiv:1006.5902.

[6] Miroslav, Bača, Koruga Petra, & Fotak Tomislav. (2011) Basic on-line handwritten signature features for personal biometric authentication. MIPRO, Proceedings of the 34th International Convention. IEEE, 2011.

[7] Kekre, H. B., & . Bharadi V. A. (2010) Gabor filter based feature vector for dynamic signature recognition. International Journal of Computer Applications 2.3(2010): 74–80.

[8] Yang, M., & Zhang, L (2010). Gabor feature based sparse representation for face recognition with gabor occlusion dictionary. In Computer Vision–ECCV 2010 (pp. 448–461). Springer Berlin Heidelberg.

[9] Alattas, Eman, & Souham Meshoul. (2011) An effective feature selection method for on- line signature based authentication. Fuzzy Systems and Knowledge Discovery (FSKD), 2011 Eighth International Conference on. Vol. 3. IEEE (2011).

[10] Al-Omari, Yazan M., Siti Norul Huda Sheikh Abdullah, & Khairuddin Omar. (2011) State-of-the- art in offline signature verification system. Pattern Analysis and Intelligent Robotics (ICPAIR), International Conference on. Vol. 1. IEEE, 2011.

[11] Kanoun, Slim, et al. (2000) Script Identification for Arabic and Latin Printed and handwritten Documents. 4th IAPR-International Workshop on Document Analysis Systems: DAS 2000.

[12] Zhang, Jianping, et al. (2014) A local information based variational model for selective image segmentation. Inverse Problems Imaging 8.1(2014): 293–320.

[13] Tan, Xiuhu, et al. "Off-line signature verification system based on dwt and common features extraction." Journal of theoretical and applied information technology 51.2 (2013).

[14] Otsu, N. (1979) A Threshold Selection Method from Gray-Level Histograms. Systems, Man and Cybernetics. IEEE Transactions on, 9 (1), 62–66.

[15] Anwar Yahy Ebrahim, Ghazali Sulong. (2014) Offline Handwritten Signature Verification Using Back Propagation Artificial Neural Network Matching Technique. 31st July. Vol. 65 No.3 © 2005–2014 JATIT & LLS. All rights reserved. 2014.

[16] Ghazali Sulong , Anwar Yahy Ebrahim , Muhammad Jehanzeb. (2014) Offline Signature Verification Using Window Based Techniques.

International Journal of Interactive Digital Media, Vol. 2(1), ISSN 2289–4098.

[17] Sulong, Ghazali, Anwar Yahy Ebrahim, and Muhammad Jehanzeb. (2014) Offline handwritten signature identification using adaptive window positioning techniques. arXiv preprint arXiv:1407.2700.

[18] Rivard D., Granger E., Sabourin R. (2013) Multi-Feature extraction and selection in writer-independent offline signature verification. International Journal on Document Analysis and Recognition 16 (1) (2013) 83–103.

[19] Anwar Yahya Ebrahim, Asmaa Shaker (2018). A New Model of Arabic Handwritten Recognition Using Combination Between DWT with Data Reduction Method. Journal of Theoretical and Applied Information Technology 96 (1992–8645), 1

[20] Sutton, Clifton D. (2005) Classification and regression trees, bagging, and boosting. Handbook of statistics 24 (2005): 303–329.

[21] Feng, Hao, and Chan Choong Wah. (2003) Online signature verification using a new extreme points warping technique. Pattern Recognition Letters 24.16 (2003): 2943–2951.

[22] Abdalla Ali A.A., Zhirkov V.F. (2009) Offline Handwritten Signature Verification Using Radon Transform and SVM/KNN Classifiers. ISSN 0136–5835. Вестник ТГТУ. 2009. Том 15. № 1. Transactions TSTU.

[23] Kisku DR., Gupta P., Sing JK. (2010) Fusion of Multiple Matchers using SVM for Offline Signature Identification. arXiv preprint arXiv: 1002.0416- arxiv.org, 2010.

[24] Bharathi R.K, Shekar. Discriminative B.H. (2014) DCT-MLP based approach for off-line signature verification. 2014 International Conference on Advances in Computing, Communications and Informatics (ICACCI), 2309–2315, (2014).

[25] M.A. Ismail and Samia Gad, (2000). Off-line Arabic Signature Recognition and Verification. PR, 2000,pp.1727–1740.

[26] AY. Ebrahim, (2017). Classification of Arabic Autograph as Genuine and Forged Through a Combination of New Attribute Extraction Techniques. Journal of University of Babylon, Pure and Applied Sciences, Vol.(25), No.(5), 2017.

[27] SM Darwish, ZH Noori, (2018). Distance and Fuzzy Classifiers Alliance: The Solution to Off- line Arabic Signature Verification System for Forensic Science. International Journal Of Artificial. Intelegence Research Vol 2, No 2, December 2018.

[28] C.ERGÜN, S. NOROZPOUR, (2019). Farsi document image recog nition system using word layout signa ture 27: 1477 – 1488T urkish Journal of Electrical Engineering & Computer. Available from: [accessed Feb 06 2021].

Chapter 7

Current State-of-the-Art of Clustering Methods for Gene Expression Data with RNA-Seq

Ismail Jamail and Ahmed Moussa

Abstract

Latest developments in high-throughput cDNA sequencing (RNA-seq) have revolutionized gene expression profiling. This analysis aims to compare the expression levels of multiple genes between two or more samples, under specific circumstances or in a specific cell to give a global picture of cellular function. Thanks to these advances, gene expression data are being generated in large throughput. One of the primary data analysis tasks for gene expression studies involves data-mining techniques such as clustering and classification. Clustering, which is an unsupervised learning technique, has been widely used as a computational tool to facilitate our understanding of gene functions and regulations involved in a biological process. Cluster analysis aims to group the large number of genes present in a sample of gene expression profile data, such that similar or related genes are in same clusters, and different or unrelated genes are in distinct ones. Classification on the other hand can be used for grouping samples based on their expression profile. There are many clustering and classification algorithms that can be applied in gene expression experiments, the most widely used are hierarchical clustering, k-means clustering and model-based clustering that depend on a model to sort out the number of clusters. Depending on the data structure, a fitting clustering method must be used. In this chapter, we present a state of art of clustering algorithms and statistical approaches for grouping similar gene expression profiles that can be applied to RNA-seq data analysis and software tools dedicated to these methods. In addition, we discuss challenges in cluster analysis, and compare the performance of height commonly used clustering methods on four different public datasets from recount2.

Keywords: clustering, classification, RNA-seq, gene expression, adjusted Rand index, machine learning, deep learning

1. Introduction

In recent years, RNA-seq based on Next generation Sequencing has become an attractive alternative for conducting quantitative analysis of gene expression. This approach offers a number of advantages compared to microarray analysis such as the discovery of novel RNA species (RNA-seq is not limited by prior knowledge of the genome of the organism, it can be used for the detection of novel transcripts), the higher sensitivity for genes expressed either at low or very high level and the unbiased approach compared to microarrays that are subject to cross-hybridization

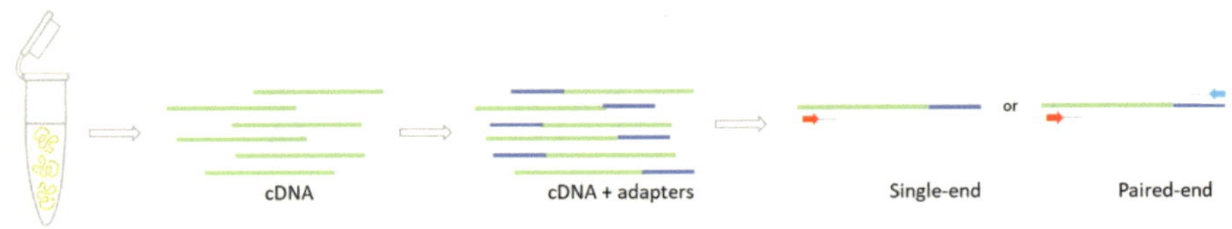

Figure 1.
RNA sequencing.

bias. Overall, RNA-seq is a better technique for many applications such as novel gene identification, differential gene expression, and splicing analysis.

The principle of RNA-seq is based on high-throughput next generation sequencing (NGS) technologies. The first step in the technique involves converting the population of RNA to be sequenced into cDNA fragments with adaptors attached to one or both ends, each molecule is then sequenced to obtain either single end short sequence reads or paired end reads [1]. These reads are stored in fastq files formats and consist of raw data for many analysis pipelines (**Figure 1**).

The primary objective of this chapter is to present algorithms for clustering gene expression data from RNA-seq. Therefore, in the first section, we will describe the different steps of the gene expression analysis workflow from preprocessing the raw reads to gene expression clustering and classification. In the second part of the chapter we will describe traditional, model-based and machine learning clustering methods for gene expression data, then we will conclude this chapter with a study for clustering samples of four public datasets from recount2, using different clustering methods and also evaluating the performance of each one using the adjusted rand index (RDI) and accuracy.

2. RNA-seq data analysis

RNA-seq has become a common tool for scientists to study the transcriptome complexity, and a convenient method for the analysis of differential gene expression. A typical RNA-seq data analysis workflow starts by preprocessing raw reads for contamination removal and quality control checks. The following step is to align the reads to a reference genome, or to make a de novo assembly if there is not any. Following the alignment, the quantification step aims to quantify aligned reads to produce a count matrix to use as entry data for Differential Expression (DE) analysis. Normalization and DE analysis normally go together as most of the methods have built-in normalization and accept only raw count matrix. For this study, we are more interested in the clustering step, we will perform Normalization of the raw counts separately and do the clustering without going through differential gene expression analysis. In the following section we describe with more details each step of the pipeline (**Figure 2**).

2.1 Preprocessing

Preprocessing raw reads consist of checking the quality of the reads, adapters trimming, removal of short reads and filtering bad quality bases. Tools like FastQC can generate a report summarizing the overall quality of the sequence information [2]. Based on this report we can determine how the quality trimming should be set up. Trimmomatic is one of many tools used to clean up the raw data. It can be used to remove adapters from the reads, trim off any low-quality bases at the ends of reads, and filter short reads that can align to multiple locations on the reference genome. Once the trimming step is done, it is a good practice to recheck the quality of the reads by rerunning FastQC.

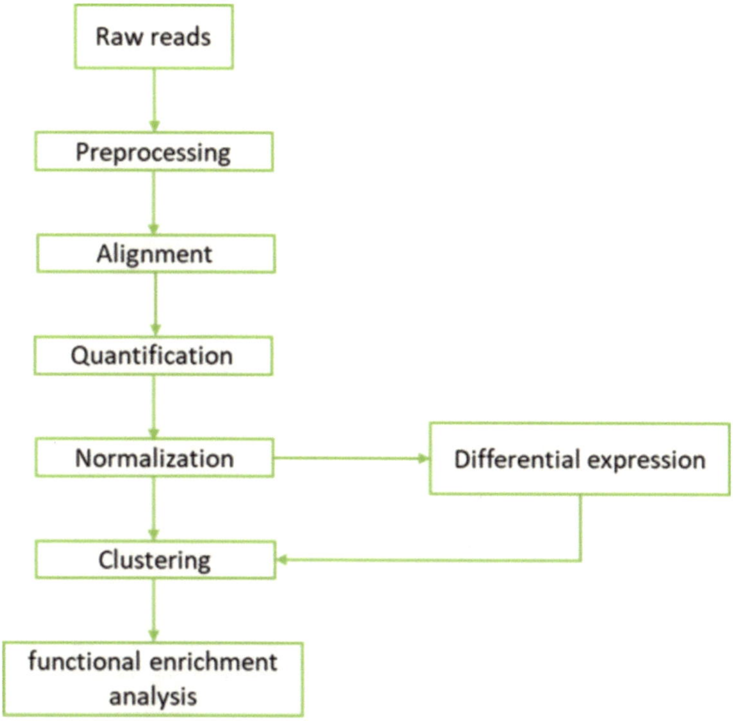

Figure 2.
RNA-seq data analysis workflow.

2.2 Alignment

Now that we have explored the quality of our raw reads, we can move on to read alignment. Read alignment is one of the first steps required for many different types of analysis. It aims to map the huge number of short RNA sequences generated by NGS instruments (reads) to a reference genome in order to identify the correct genomic loci from which the read originated. In RNA-seq, alignment is a major step for the calculation of transcript or gene expression levels; several splice-aware alignment methods have been developed for RNA-seq experiments such as STAR, HISAT2 or TopHat. These aligners are designed to specifically address many of the challenges of RNA-seq data mapping using a strategy to account for spliced alignments [3–5].

2.3 Quantification

Quantification of gene expression is to count the number of reads that map to each gene using methods such as HTSeq-count, FeatureCounts or kallisto [6–8]. This step is crucial if we want to do a gene differential expression analysis, which means to identify genes (or transcripts), if any, that have a statistically significant difference in abundance across the experimental groups or conditions.

2.4 Normalization

The read counts generated in the quantification step need to be normalized to make accurate comparisons of gene expression between samples or when doing an exploratory data analysis. Several normalization methods are used for this purpose such as CPM (counts per million), TPM (transcripts per kilobase million), RPKM/FPKM

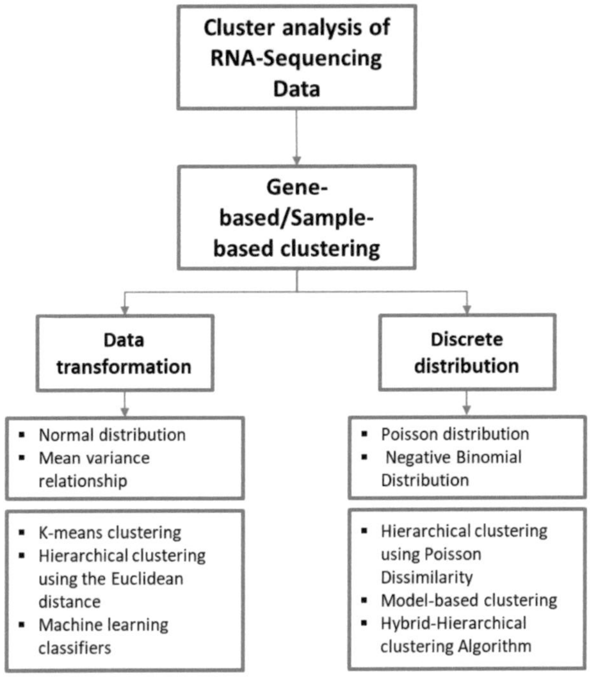

Figure 3.
Cluster analysis of RNA-sequencing data.

(reads/fragments per kilobase of exon per million reads/fragments mapped), DESeq2's median of ratios and EdgeR's trimmed mean of M values (TMM) [9].

2.5 Clustering

Cluster analysis techniques have proven to be helpful to understand gene expression data by uncovering unknown relationships among genes and unveiling different subtypes of diseases when it comes to clustering biological samples [10]. In the following section, we present methods for sample-based and gene-based clustering, starting with traditional methods used after data transformation then model-based clustering for data generated using a combination of probability distributions (**Figure 3**).

3. Clustering methods for gene expression data

3.1 Data transformation methods

Traditional clustering algorithms like hierarchical clustering and k-means cannot be directly applied to RNA-seq count data, to apply these methods for cluster analysis of RNA-seq data, that tend to follow an over-dispersed Poisson or negative binomial distribution, we need to transform the data in order to have a distribution closer to the normal distribution. In the following section, we present popular methods for data transformation:

- Logarithmic, widely used method to deal with skewed data in many research domains, often used to reduce the variability of the data and make the data conform more closely to the normal distribution. However, it was demonstrated in [11], that in most circumstances the log transformation does not help make data less variable or more normal and may, in some circumstances, make data more variable and more skewed.

- Variance stabilizing transformation: This method was used to transform microarray data to stabilize the asymptotic variance over the full range of the data [12].

- Eight data transformations (r, r2, rv, rv2, l, l2, lv, and lv2) for RNA-seq data analysis were proposed in [13], these methods deal with the two common properties when it come to the count matrix generated in the quantification step, Sparsity and Skewness; Sparsity means that many counts in the count matrix are zero. Skewness means that the histogram of all counts in the count matrix is usually skewed.

3.2 Clustering methods based on normal distribution

3.2.1 Hierarchical methods

Hierarchical clustering method is the most popular method for gene expression data analysis. In hierarchical clustering, genes with similar expression patterns are grouped together and are connected by a series of branches (clustering tree or dendrogram). Experiments with similar expression profiles can also be grouped together using the same method. This clustering technique is divided into two types: agglomerative and divisive. In an agglomerative or bottom-up clustering method each observation is assigned to its own cluster. In a comparative study on Cancer data [14], three variants of Hierarchical Clustering Algorithms (HCAs): Single-Linkage (SL), Average-Linkage (AL) and Complete-Linkage (CL) with 12 distance measure have been used to cluster RNA-seq Samples. The same methods will be used in our study along with hierarchical clustering with Poisson distribution [15].

3.2.2 k-medoids

K-medoids is a partitional clustering algorithm proposed in 1987 by Kaufman and Rousseeuw. It is a variant of the K-means algorithm that is less sensitive to noise and outliers because it uses medoids as cluster centers instead of means that are easily influenced by extreme values. Medoids are the most centrally located objects of the clusters, with a minimum sum of distances to other points. After searching for k representative objects in a data set, the algorithm which is called Partitioning Around Medoids (PAM) assigns each object to the closest medoid in order to create clusters. Like in k-means the number of classes to be generated needs to be specified.

3.3 Model-based clustering

Yaqing Si et al. described a number of Model based clustering methods for RNA-seq data in their paper [16], these methods assume that data are generated by a mixture of probability distributions: Poisson distribution when only technical replicates are used and Negative binomial distribution when working with biological replicates. The first method they proposed is a model-based clustering method with the expectation-maximization algorithm (MB-EM) for clustering RNA-seq gene expression profile. The expectation-maximization algorithm is widely used in many computational biology applications, the authors in [17] explain how this algorithm works and when it is used. The second method is an initialization algorithm for cluster centers, the idea behind this method is to randomly choose one cluster center and then gradually add centers by selecting genes based on the distance between each gene and each of the selected centers. Two other stochastic algorithms have been proposed in this paper, a stochastic version of the expectation-maximization algorithm and a classification expectation maximization algorithm with simulated annealing. The last method in this paper is a model-Based Hybrid-Hierarchical Clustering Algorithm, it does not require to pre-specify the number of clusters to be generated as it is required by the previous methods. The authors propose to use agglomerative clustering starting with k0 clusters to speed up the calculation, then, it repeatedly identifies the two clusters that are closest together and merges the two most similar clusters. This method was called hybrid because it combines two steps: Obtaining the initial K0 clusters using one of the previous described algorithms then agglomerative clustering to build the hierarchical tree.

3.4 Classification and clustering algorithms of machine learning for RNA-seq data

Classification in machine learning is a supervised learning approach in which the algorithm learns from the data given to it and makes new observations, then applies the conclusions to new data. Clustering on the other hand is an unsupervised learning problem for grouping unlabeled features. The learning algorithm that learns the model from the training data and maps the input data to a specific class is called classifier, in the following section, we briefly present three widely used classifiers for grouping RNA-seq data.

- Random forests (RF): an ensemble method that trains a large number of individual decision trees, each tree gives a class prediction, the category that wins the majority votes is used as the final decision of the random forest model. The algorithm can perform both classification and regression tasks and has better accuracy among current algorithms.

- Support Vector Machine (SVM): one of the most popular supervised learning models, used for both classification and regression, the data points are separated using an optimal hyperplane or a set of hyperplanes in a multidimensional space with the maximum possible margin between support vectors.

- Poisson linear discriminant analysis: an approach used for the classification and clustering of RNA-seq data using a Poisson log linear model [15].

To test these algorithms, we used MLSeq (Machine learning interface for RNA-sequencing data) which is an R package including more than 80 machine learning algorithms and a pipeline to classify RNA-seq data including normalization, filtering and transformation steps [18].

3.5 Clustering with deep learning

Deep learning is also a technique that can be used to learn better data representation of high-dimensional data. The two recently published surveys [19, 20] present a taxonomy of existing deep clustering algorithms, by describing the different Neural Network Architecture that exists for feature representation, clustering loss function and Performance Evaluation Metrics for Deep Clustering. In [20], the authors categorize current deep clustering models into following three categories:

- Auto-Encoders Based Deep Clustering
- CDNN-Based Deep Clustering (feed-forward networks trained only by specific clustering)
- Generative Adversarial Network (GAN)

These approaches are already used in the analysis of RNA-seq data, for example, an unsupervised deep embedding algorithm that clusters single cell (scRNA-seq) data was proposed in [21], another paper use a Lasso model and a multilayer feed-forward artificial neural network to analyze RNA-Seq gene expression data [22]. In [23], the authors used a Deep Neural Network model from the R package h2o for cancer data classification and in [24], ladder networks were used for gene expression classification.

4. Clustering algorithms and software packages/tools corresponding to the algorithms

Clustering algorithms and software packages corresponding to the algorithms are shown in **Table 1**.

Methods	Implementation in R
Hierarchical clustering	hclust() function in "stats"
k-means	"cluster", "factoextra"
k-medoids	"cluster", "factoextra"
SOM	"kohonen"
Model-based clustering with the expectation-maximization algorithm (MB-EM). Stochastic version of the expectation-maximization algorithms (Deterministic annealing (DA) algorithm). Classification expectation maximization (CEM) algorithm with simulated annealing (SA).	"MBCluster.Seq"
Machine learning algorithms	"MLSeq"

Table 1.
Clustering algorithms and software packages corresponding to the algorithms.

5. Clustering of public RNA-seq data from recount2

Recount2 is a multi-experiment resource of analysis-ready RNA-seq gene and exon count datasets. It contains 2041 different studies and over 70,000 human RNA-seq [25]. We selected for our study four different datasets based on the number of samples and the number of classes. We then performed sample-based clustering on each dataset and compared the results to the classes in the phenotype table in recount2 to evaluate the performance of each method. The methods used to classify the data are 3 subtypes of the hierarchical clustering with the Euclidean distance, hierarchical clustering with a Poisson model and k-medoids.

5.1 Datasets

Description of the four datasets from recount2 is shown in **Table 2**.

5.2 Adjusted Rand Index

There are several similarity measures for cluster evaluation, we chose to work with the adjusted Rand index which is the corrected-for-chance version of the Rand index. It is a measure used in data clustering to evaluate the performance of a clustering method, by comparing the results of a clustering algorithm against known classes from external criteria [26]. In our study, we performed different sample-based classification method on four different datasets, after that, we compared the results to the class labels we associated to each sample based on the field "characterization of the samples" in the phenotype table in recount2, then we used the ARI for cluster validation.

5.3 Standard deviation

Figures 4–6 are examples to show the standard deviation of the transformed data, across samples, against the mean, using the shifted logarithm transformation, the regularized log transformation and the variance stabilizing transformation.

Dataset (accession)	Number of samples	Number of classes	Classes
SRP032789	20	4	17 breast tumor samples of three different subtypes: • TNBC. • Non-TNBC. • HER2-positive.
SRP049097	54	4	3 subtypes of Leiomyosarcoma: • 8 LMS cases from subtype I • 6 cases from subtype II • 3 cases from subtype III • 7 cases of normal tissues
SRP042620	168	6	• 28 breast cancer cell lines. • 42 Triple Negative Breast Cancer (TNBC) primary tumors. • 42 Estrogen Receptor Positive (ER+) and HER2 Negative Breast Cancer primary tumors. • 30 uninvolved breast tissue samples that were adjacent to ER+ primary tumors. • 5 breast tissue samples from reduction mammoplasty procedures performed on patients with no known cancer. • 21 uninvolved breast tissue samples that were adjacent to TNBC primary tumors.
SRP044668	94	3	• 39 contrast-enhancing glioma core samples. • 36 non-enhancing FLAIR glioma margin samples. • 17 non-neoplastic brain tissue samples.

Table 2.
Description of the four datasets from recount2.

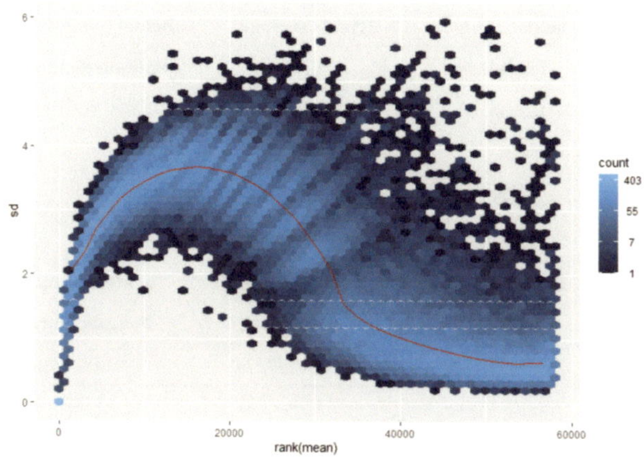

Figure 4.
Standard deviation of the transformed data using the shifted logarithm transformation.

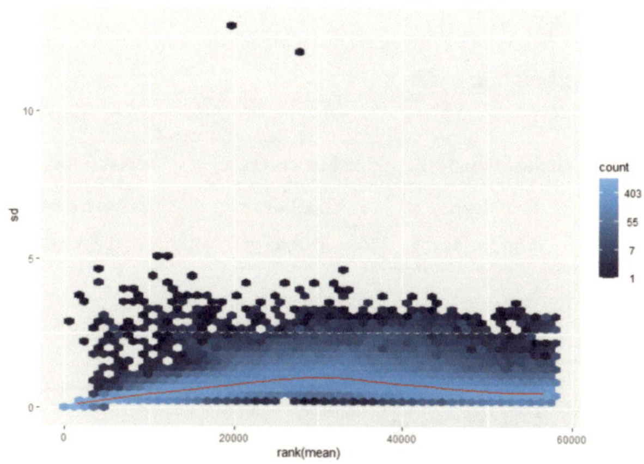

Figure 5.
Standard deviation of the transformed data using the regularized log transformation.

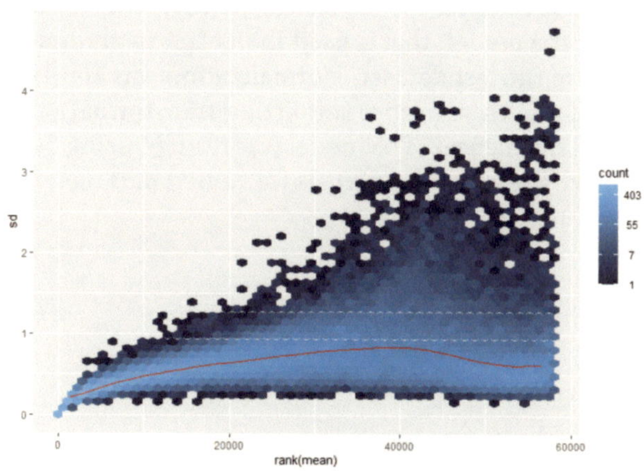

Figure 6.
Standard deviation of the transformed data using the variance stabilizing transformation.

hclus (complete)	hclust (single)	hclust (average)	hclust (complete)	k-medoids
Euclidean	Euclidean	Euclidean	Poisson distance	Euclidean
0.4146015	0.3818763	0.4146015	0.4146015	0.6798897

Table 3.
Performance of clustering methods (SRP032789).

hclus (complete)	hclust (single)	hclust (average)	hclust (complete)	k-medoids
Euclidean	Euclidean	Euclidean	Poisson distance	Euclidean
0.02880412	−0.003409256	0.0005777741	0.1874828	0.2791547

Table 4.
Performance of clustering methods (SRP049097).

hclus (complete)	hclust (single)	hclust (average)	hclust (complete)	k-medoids
Euclidean	Euclidean	Euclidean	Poisson distance	Euclidean
0.1944569	0.005551586	0.1285448	0.1468464	0.2579758

Table 5.
Performance of clustering methods (SRP042620).

hclus (complete)	hclust (single)	hclust (average)	hclust (complete)	k-medoids
Euclidean	Euclidean	Euclidean	Poisson distance	Euclidean
0.2379903	−0.007755123	0.399417	0.2657942	0.3771837

Table 6.
Performance of clustering methods (SRP044668).

5.4 Machine learning classification

Three widely used machine learning algorithms were used for the classification of the four datasets, Random forests, support vector machine and Poisson linear discriminant analysis. To perform this analysis, we first split the data into two parts as training and test sets, with 70% of samples for the training dataset, and the remaining 30% samples for the testing dataset, the training set is used to fit the parameters of the model, that is used thereafter to predict the responses for the observations in the test dataset. Normalization was applied with Deseq median ratio method and the variance stabilizing transformation was applied for the normalization of the dataset. The model was trained using 5-fold cross validation repeated 2 times. The number of levels for tuning parameters is set to 10.

5.5 Results

Classifier	Accuracy	Kappa
rf	1	1
SVM	0.6667	0.5
PLDA	1	1

Table 7.
Classification results for SRP032789 data.

Classifier	Accuracy	Kappa
rf	0.8235	0.765
SVM	0.7647	0.6909
PLDA	0.7647	0.6866

Table 8.
Classification results for SRP049097 data.

Classifier	Accuracy	Kappa
rf	0.9412	0.9249
SVM	0.5882	0.4685
PLDA	0.7843	0.7267

Table 9.
Classification results for SRP042620 data.

Classifier	Accuracy	Kappa
rf	0.8214	0.7271
SVM	0.6786	0.5218
PLDA	0.7143	0.5573

Table 10.
Classification results for SRP044668 data.

5.6 Computational time

All experiments are performed on a machine with 16 GB RAM, 1024 GB hard disk running with a windows operating system and MLSeq R package.

	SRP032789	SRP049097	SRP042620	SRP044668
rf	176.67	781.31	4234.89	1412.19
SVM	1080.92	2333.52	6645.21	1597.89
PLDA	31.45	60.93	234.98	72.66

Table 11.
Computational time in seconds.

6. Discussion and conclusion

Clustering the samples of the three datasets to the sub-classes defined in the phenotype table of recounts2 was not easy. We first tried to visualize the separation between the subtypes using principal component analysis (**Figures** 7 and **8–10**), then using 4 variants of the hierarchical clustering and k-medoids we classified the samples of each dataset (**Figures 11** and **12** show the hierarchical clustering plots of the dataset SRP032789). The performance of the 5 methods was different depending on the dataset (**Tables 3–5**), making it impossible to make a general system of recommendation. However, we can see that the k-medoid method has relatively

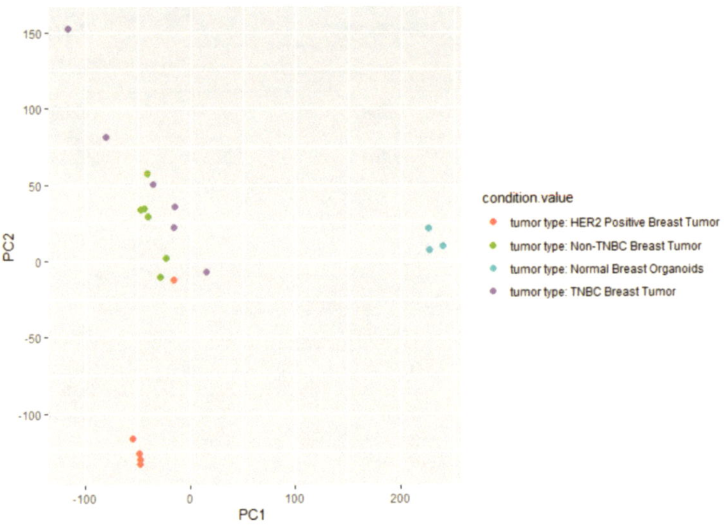

Figure 7.
PCA of data from the study SRP032789.

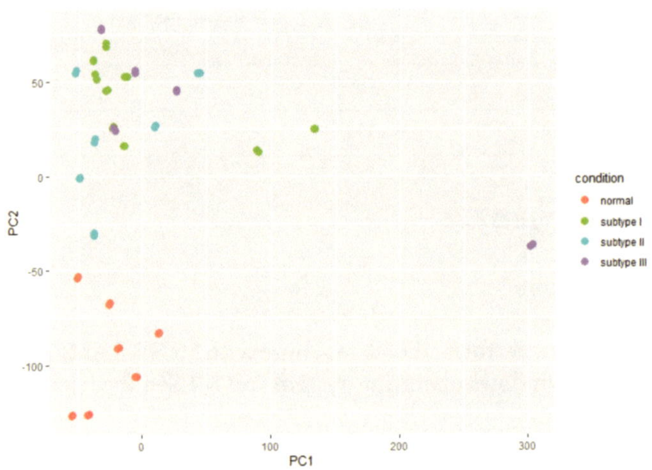

Figure 8.
PCA of the data from the study SRP049097.

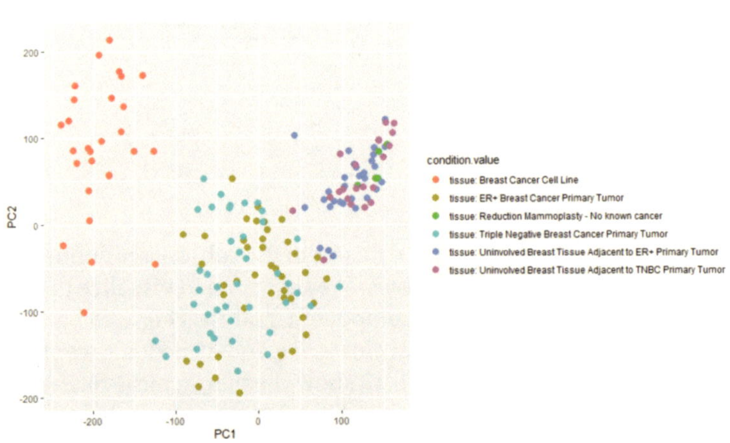

Figure 9.
PCA of the data from the study SRP042620.

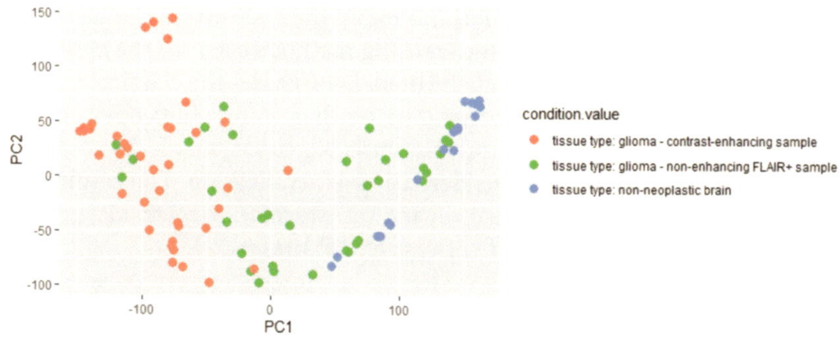

Figure 10.
PCA of the data from the study SRP044668.

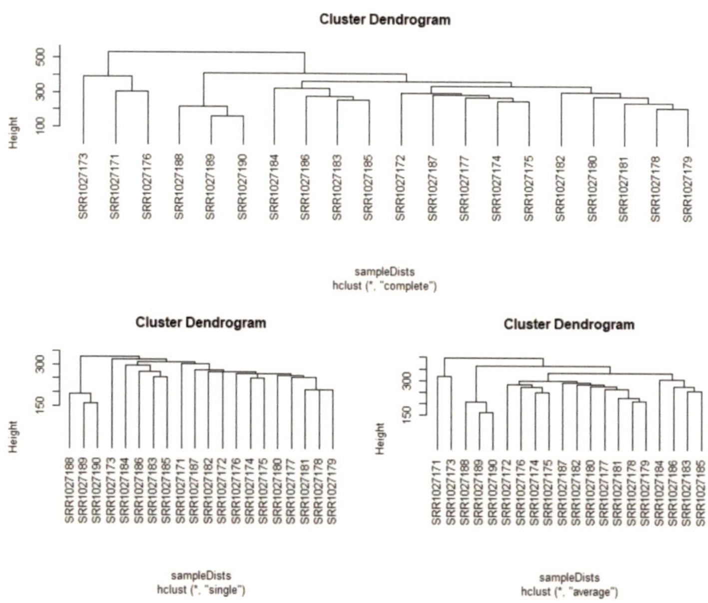

Figure 11.
Dendrograms obtained for the dataset from SRP032789 study using three variants of the hierarchical clustering method with the Euclidean distance.

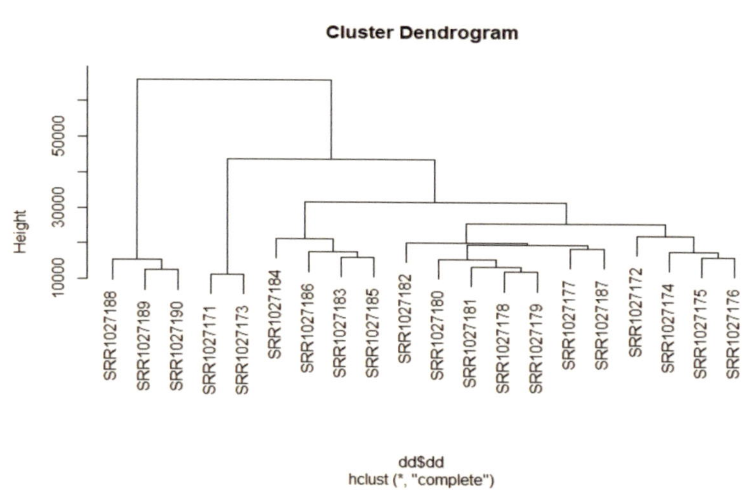

Figure 12.
Dendrograms obtained for the dataset from the study SRP032789 using the hierarchical clustering method with the Poisson distance.

better performance than the other methods for all the datasets. In the second part of the study, we compared a few machine learning methods used for the classification of RNA-seq data. The performance of the models surpasses the classical methods used before, also RF and PLDA performed better than SVM which does not perform very well when the data set is large and has noise. Note that the model accuracies given in this study should not be considered as a generalization. The results can depend on several criteria: normalization and transformation methods, gene-wise overdispersions, outliers, number of classes etc. (**Tables 6–11**).

Conflict of interest

The authors declare no conflict of interest.

Author details

Ismail Jamail* and Ahmed Moussa
System and Data Engineering Team - SDET, Tangier, Morocco

*Address all correspondence to: jamail@ensat.ac.ma

References

[1] Wang Z, Gerstein M, Snyder M. RNA-Seq: a revolutionary tool for transcriptomics. Nat Rev Genet [Internet]. 2009 Jan; 10(1):57-63. Available from: http://dx.doi.org/10.1038/nrg2484

[2] Andrews S. (2010). FastQC: a quality control tool for high throughput sequence data. Available online at: http://www.bioinformatics.babraham.ac.uk/projects/fastqc

[3] Dobin A, Davis CA, Schlesinger F, Drenkow J, Zaleski C, Jha S, et al. STAR: ultrafast universal RNA-seq aligner. Bioinformatics [Internet]. 2012 Oct 25;29(1):15-21. Available from: http://dx.doi.org/10.1093/bioinformatics/bts635

[4] Kim D, Langmead B, Salzberg SL. HISAT: a fast spliced aligner with low memory requirements. Nat Methods [Internet]. 2015 Mar 9;12(4):357-60. Available from: http://dx.doi.org/10.1038/nmeth.3317

[5] Trapnell C, Pachter L, Salzberg SL. TopHat: discovering splice junctions with RNA-Seq. Bioinformatics [Internet]. 2009 Mar 16;25(9):1105-11. Available from: http://dx.doi.org/10.1093/bioinformatics/btp120

[6] Anders S, Pyl PT, Huber W. HTSeq--a Python framework to work with high-throughput sequencing data. Bioinformatics [Internet]. 2014 Sep 25;31(2):166-9. Available from: http://dx.doi.org/10.1093/bioinformatics/btu638

[7] Liao Y, Smyth GK, Shi W. featureCounts: an efficient general purpose program for assigning sequence reads to genomic features. Bioinformatics [Internet]. 2013 Nov 13;30(7):923-30. Available from: http://dx.doi.org/10.1093/bioinformatics/btt656

[8] Bray NL, Pimentel H, Melsted P, Pachter L. Near-optimal probabilistic RNA-seq quantification. Nat Biotechnol [Internet]. 2016 Apr 4;34(5):525-7. Available from: http://dx.doi.org/10.1038/nbt.3519

[9] Li P, Piao Y, Shon HS, Ryu KH. Comparing the normalization methods for the differential analysis of Illumina high-throughput RNA-Seq data. BMC Bioinformatics [Internet]. 2015 Oct 28;16(1). Available from: http://dx.doi.org/10.1186/s12859-015-0778-7

[10] Vidman L, Källberg D, Rydén P. Cluster analysis on high dimensional RNA-seq data with applications to cancer research - An evaluation study. Nazarov PV, editor. PLoS ONE [Internet]. 2019 Dec 5;14(12):e0219102. Available from: http://dx.doi.org/10.1371/journal.pone.0219102

[11] Feng, Changyong et al. "Log-transformation and its implications for data analysis." Shanghai archives of psychiatry vol. 26,2 (2014): 105-9. doi:10.3969/j.issn.1002-0829.2014.02.009

[12] Durbin BP, Hardin JS, Hawkins DM, Rocke DM. A variance-stabilizing transformation for gene-expression microarray data. Bioinformatics [Internet]. 2002 Jul 1;18(Suppl 1):S105-10. Available from: http://dx.doi.org/10.1093/bioinformatics/18.suppl_1.S105

[13] Zhang Z, Yu D, Seo M, Hersh CP, Weiss ST, Qiu W. Novel Data Transformations for RNA-seq Differential Expression Analysis. Sci Rep [Internet]. 2019 Mar 18;9(1). Available from: http://dx.doi.org/10.1038/s41598-019-41315-w

[14] Jaskowiak PA, Costa IG, Campello RJGB. Clustering of RNA-Seq samples: Comparison study on

cancer data. Methods [Internet]. 2018 Jan;132:42-9. Available from: http://dx.doi.org/10.1016/j.ymeth.2017.07.023

[15] Witten DM. Classification and clustering of sequencing data using a Poisson model. Ann Appl Stat [Internet]. 2011 Dec;5(4):2493-518. Available from: http://dx.doi.org/10.1214/11-AOAS493

[16] Si Y, Liu P, Li P, Brutnell TP. Model-based clustering for RNA-seq data. Bioinformatics [Internet]. 2013 Nov 4;30(2):197-205. Available from: http://dx.doi.org/10.1093/bioinformatics/btt632

[17] Do CB, Batzoglou S. What is the expectation maximization algorithm? Nat Biotechnol [Internet]. 2008 Aug;26(8):897-9. Available from: http://dx.doi.org/10.1038/nbt1406

[18] Goksuluk D, Zararsiz G, Korkmaz S, Eldem V, Zararsiz GE, Ozcetin E, et al. MLSeq: Machine learning interface for RNA-sequencing data. Computer Methods and Programs in Biomedicine [Internet]. 2019 Jul;175:223-31. Available from: http://dx.doi.org/10.1016/j.cmpb.2019.04.007

[19] Aljalbout, Elie et al. Clustering with Deep Learning: Taxonomy and New Methods. *ArXiv* abs/1801.07648. 2018

[20] Min E, Guo X, Liu Q, Zhang G, Cui J, Long J. A Survey of Clustering With Deep Learning: From the Perspective of Network Architecture. IEEE Access [Internet]. 2018;6:39501-14. Available from: http://dx.doi.org/10.1109/ACCESS.2018.2855437

[21] Li X, Wang K, Lyu Y, Pan H, Zhang J, Stambolian D, et al. Deep learning enables accurate clustering with batch effect removal in single-cell RNA-seq analysis. Nat Commun [Internet]. 2020 May 11;11(1). Available from: http://dx.doi.org/10.1038/s41467-020-15851-3

[22] Urda D, Montes-Torres J, Moreno F, Franco L, Jerez JM. Deep Learning to Analyze RNA-Seq Gene Expression Data. In: Advances in Computational Intelligence [Internet]. Springer International Publishing; 2017. p. 50-9. Available from: http://dx.doi.org/10.1007/978-3-319-59147-6_5

[23] Sharma A, Rani R. An Optimized Framework for Cancer Classification Using Deep Learning and Genetic Algorithm. j med imaging hlth inform [Internet]. 2017 Dec 1;7(8):1851-6. Available from: http://dx.doi.org/10.1166/jmihi.2017.2266

[24] Golcuk G, Tuncel MA, Canakoglu A. Exploiting Ladder Networks for Gene Expression Classification. In: Bioinformatics and Biomedical Engineering [Internet]. Springer International Publishing; 2018. p. 270-8. Available from: http://dx.doi.org/10.1007/978-3-319-78723-7_23

[25] Collado-Torres L, Nellore A, Kammers K, Ellis SE, Taub MA, Hansen KD, et al. Reproducible RNA-seq analysis using recount2. Nat Biotechnol [Internet]. 2017 Apr;35(4):319-21. Available from: http://dx.doi.org/10.1038/nbt.3838

[26] Santos JM, Embrechts M. On the Use of the Adjusted Rand Index as a Metric for Evaluating Supervised Classification. In: Artificial Neural Networks – ICANN 2009 [Internet]. Springer Berlin Heidelberg; 2009. p. 175-84. Available from: http://dx.doi.org/10.1007/978-3-642-04277-5_18